THE BRIDGEWATER TREATISES

ON THE POWER WISDOM AND GOODNESS OF GOD
AS MANIFESTED IN THE CREATION

———

TREATISE VI

GEOLOGY AND MINERALOGY CONSIDERED WITH REFERENCE

TO NATURAL THEOLOGY

BY THE REV. WILLIAM BUCKLAND, D.D.

IN TWO VOLUMES

VOL II

THOU LORD IN THE BEGINNING HAST LAID THE FOUNDATION OF THE EARTH.
PSALM CII. 25.

" Let us take a Survey of the principal Fabrick, viz. the Terraqueous Globe itself; a most stupendous work in every particular of it, which doth no less aggrandize its Maker than every curious complete work doth its Workman. Let us cast our eyes here and there, let us ransack all the Globe, let us with the greatest accuracy inspect every part thereof, search out the inmost secrets of any of the creatures, let us examine them with all our gauges, measure them with our nicest rules, pry into them with our microscopes and most exquisite instruments, still we find them to bear testimony to their infinite Workman."

DERHAM'S PHYSICO-THEOLOGY, BOOK II. P. 38.

" Could the body of the whole Earth - - be submitted to the Examination of our Senses, were it' not too big and disproportioned for our Enquiries, too unwieldly for the Management of the Eye and Hand, there is no question but it would appear to us as curious and well-contrived a frame as that of an human body. We should see the same Concatenation and Subserviency, the same Necessity and Usefulness, the same Beauty and Harmony in all and every of its Parts, as what we discover in the Body of every single Animal."

SPECTATOR, NO. 543.

GEOLOGY AND MINERALOGY

CONSIDERED WITH REFERENCE TO

NATURAL THEOLOGY

BY

THE REV. WILLIAM BUCKLAND, D.D.

CANON OF CHRIST CHURCH AND READER IN GEOLOGY AND MINERALOGY
IN THE UNIVERSITY OF OXFORD

VOL II

ALDI

DISCIP.

ANGLVS

LONDON
WILLIAM PICKERING
1836

C. WHITTINGHAM, TOOKS COURT, CHANCERY LANE.

LIST OF ENGRAVINGS.

* Besides 120 figures of Plants and Animals, this Plate represents 30 kinds of Sedimentary Deposits, and 8 varieties of Unstratified Rocks; it also shows the dispositions of intruded Dykes, Metalliferous Veins, and Faults.

Total number of Plates 87. Total number of figures 705.

EXPLANATION OF THE PLATES.

.

*Introductory Notice, and Description of the Geological
Phenomena illustrated by Plate I.*

PLATE 1,

Is an imaginary Section constructed to express, by the insertion of names, and colours, the relative positions of the
most important classes, both of unstratified and stratified
rocks, as far as they have yet been ascertained. It is
founded on many series of accurate observations, on several
lines taken across Europe, between the British islands and
the Mediterranean Sea. Although no single straight line
exhibits every formation complete in the full order of
succession here represented, no fact is inserted for which
authority cannot be found. The near approximation of
this synoptic representation to the facts exhibited by an
actual section, may be estimated by comparing it with
the admirable section across Europe, published by Mr.
Conybeare in the Report of the Proceedings of the British
Association for the Advancement of Science, 1832, and
with his sections of England, in Phillips and Conybeare's
Geology of England and Wales.

The chief merit of the above Section is due to the
Talents of Mr. Thomas Webster; it is founded on a more
simple section which has for several years been used by
him in his lectures, and which exhibits the relations of
the Granitic and Volcanic rocks to the stratified formations,

G. II.   B

and to one another, more intelligibly than I have ever seen
expressed elsewhere. This original drawing by Mr. Webster has formed the basis of the present enlarged and
improved section, into which many important additions
have been introduced by the joint suggestions of Mr. Webster and myself. The selection and arrangement of the
animals and plants is my own; they have been drawn and
engraved (together with a large proportion of the wood-
cuts) by Mr. J. Fisher, of St. Clements, Oxford.

For facility of reference, I have numbered the principal groups of stratified rocks represented in the section,
according to their most usual order of succession; and
I have designated by letters the crystalline or unstratified
rocks, and the injected masses and dykes, as well as the
metallic veins, and lines of fracture, producing dislocations
or faults. The crowded condition in which all the Phenomena represented in this section, are set together, does
not admit of the use of accurate relative proportions,
between the stratified rocks and the intruded masses, veins,
and dykes by which they are intersected. The adoption
of false proportions is, however, unavoidable in these cases;
because the veins and dykes would be invisible, unless
expressed on a highly exaggerated scale. The scale of
height throughout the whole section is also infinitely greater
than that of breadth. The plants and animals also are
figured on no uniform scale.

The extent of the different formations represented in this
section, taking their average width as they occur in Europe,
would occupy a breadth of five or six hundred miles. A
scale of heights, at all approaching to this scale of breadth,
would render the whole almost invisible. The same cause
makes it also impossible to express correctly the effect of
vallies of *denudation*, which are often excavated through
strata of one formation into those of another subjacent
formation.

As it would encumber the section to express *Diluvium*, wherever it is present, it is introduced in one place only, which shews its age to be more recent than the newest of the Tertiary strata; it is found also lodged indiscriminately upon the surface of rocks of every formation.

Granite.

In our early Chapters we have considered the Theory which refers unstratified rocks to an igneous Origin, to be that which is most consistent with all the known Phenomena of Geology, and the facts represented in the Section now before us are more consistent with the Postulates of this Hypothesis, than with those of any other that has hitherto been proposed. I have, therefore, felt it indispensable to adopt its language, as affording the only terms by which the facts under consideration can be adequately described.

Assuming that Fire and Water have been the two great Agents employed in reducing the surface of the globe to its actual condition, we see, in repeated operations of these agents, causes adequate to the production of those irregular Elevations and Depressions of the fundamental Rocks of the Granitic series, which are delineated in the lower Region of our Section, as forming the basis of the entire Superstructure of stratified Rocks.

Near the right extremity of this Section, the undulating surface of the fundamental Granite (a. 5. a. 6. a. 7. a. 8.) is represented as being, for the most part, beneath the level of the Sea.

On the left extremity of the Section (a. 1. a. 2, a. 3.) the Granite is elevated into one of those lofty Alpine ridges, which have affected, by their upward movement, the entire series of stratified Rocks.

Corresponding formations of Primary and Transition

Strata, are represented as occurring on each side of this elevated Granite, which is supposed to have broken through,
and to have carried up with it to their present elevated and
highly inclined position, strata that were once continuous
and nearly horizontal.*

The general history of Elevation appears to be, that
mountain chains of various extent, and various directions,
have been formed at irregular intervals, during the deposition of stratified rocks of every age; and that Granite had,
in many cases, acquired a state of solidity before the
period of its elevation.

Within the primary Granite, we find other forms of Granitic matter, (a. 9.) which appear to have been intruded in a
state of fusion, not only into fissures of the older Granite,
but frequently also into the Primary stratified rocks in contact with it, and occasionally into strata of the Transition
and Secondary series, (a. 10. a. 11.) these Granitic injections were probably in many cases, contemporaneous with
the elevation of the rocks they intersect; they usually assume the Condition of Veins, terminating upwards in small
branches; and vary in dimensions, from less than an inch,
to an indefinite width. The direction of these veins is very
irregular: they sometimes traverse the Primary strata at
right angles to their planes of stratification, at other times
they are protruded in a direction parallel to these planes,
and assume the form of beds. Some of the relations of
these Granitic Veins to the rocks intersected by them are
represented at the left extremity of the Section. (a. 9.)†.

---

* Cases of Granite, thus elevated at a period posterior to the
deposition of Tertiary Strata, occur in the Eastern Alps, where 'the
Transition, Secondary, and Tertiary strata have all partaken of the
same elevation which raised the central axis of the crystalline Granitic rocks. See Geol. Trans. N. S. Vol. III. Pl. 36. Fig. 1.

† In the Granite at the right extremity of the Section, the granitic veins are omitted, because their insertion would interfere with

A. 10. represents a dyke and protruded mass of Granite, intersecting and overlying stratified rocks of the Primary and Transition series. A. 11. represents the rare case of Granite intersecting Red Sandstone, Oolite, and Chalk.*

Sienite, Porphyry, Serpentine, Greenstone.

Closely allied to Granitic Veins, is a second series of irregularly injected rocks, composed of Sienite, Porphyry, Serpentine, and Green Stone (b. c. d. e.) which traverse the Primary and Transition formations, and the lower regions of the Secondary strata; not only intersecting them in various directions, but often forming also overlying masses, in places where these veins have terminated by overflowing at the surface, (b'. c'. d'. e'.) The crystalline rocks of this series, present so many modifications of their ingredients, that numerous varieties of Sienite, Porphyry, and Greenstone occur frequently in the products of Eruptions from a single vent.

The scale of our Section admits not of an accurate representation of the relations between many of these intruded rocks, and the strata they intersect; they are all placed, as

the representation of the injections of Basaltic and Volcanic matter which that portion of the section is intended to illustrate.

* An example of the rare Phenomenon of Granite intruded into the Chalk formation, in the hill of St. Martin, near Pont de la Fou in the Pyrenees, is described by M. Dufrenoy in the Bulletin de la Société Géologique de France, Tom. 2. page 73.

At Weinböhla, near Meissen in Saxony, Prof. Weiss has ascertained the presence of Sienite above strata of Chalk; and Prof. Nauman states, that, near Oberau, Cretaceous rocks are covered by Granite, and that near Zscheila and Neiderfehre, the Cretaceous rocks rest horizontally on Granite; at both these places the Limestone and Granite are entangled in each other, and irregular portions and veins of hard Limestone, with green grains and cretaceous fossils, are here and there imbedded in the Granite.

De la Beche. Geol. Manual. 3rd Edit. p. 295.

if they had been injected, either at the time of, or after the elevation of all the strata, and had produced but little disturbance in the rocks through which they are protruded. It should however be understood, distinctly, that some Injections may have preceded the elevation of Strata to their present height, and that numerous and successive elevations and injections, attended by various degrees of fracture and disturbance, have prevailed in various localities during all periods, and throughout all formations; from the first upraising of the earliest Primary rocks, to the most recent movements produced by existing Volcanoes. M. Elie de Beaumont has discovered probable evidence of no less than twelve periods of elevation, affecting the strata of Europe.

Examples of the fractures and dislocations attending these movements, and producing faults, are represented in our Section by the lines designated by the letter l. Some of these fractures do not reach to the present surface, as they affected the lower beds at periods anterior to the deposition of more recent strata, which cover unconformably the summits of the earlier fractures. (See l. l¹. l². l³. l⁶. l⁷.)

Basalt.

A third series of Igneous Rocks is that which has formed dykes, and masses of Basalt and Trap, intruded into, and overlying formations of all ages, from the earliest Granites to the most recent Tertiary Strata. These basaltic rocks sometimes occur as Beds, nearly parallel to the strata, into which they are protruded, after the manner represented in the carboniferous Limestone of our Section, f. 2. More frequently they overspread the surface like expanded sheets of Lava. Our Section gives examples of Trap under all these circumstances. At f. 1. it intersects and overlies Primary strata; at f. 2. f. 3. f. 4. f. 5. it stands in similar relations to Transition and Secondary strata; f. 6. represents an example of an extensive eruption of Basaltic matter, over Chalk and Tertiary strata, accompanied by an intrusion of vast irregular masses of the same materials into the body of the subjacent Primary and Transition rocks.

f. 7. represents strata of columnar Basalt, immediately beneath streams of cellular Lava, in regions occupied also by craters of extinct Volcanos, f. 8. represents similar beds of columnar lava in the vicinity of active Volcanos.

Trachyte and Lava.

The fourth and last class of intruded rocks, is that of modern volcanic Porphyries, Trachytes,* and Lavas. The undeniable igneous origin of rocks of this class forms the strongest ground-work of our arguments, in favour of the igneous formation of the older unstratified and crystalline rocks; and their varied recent products, around the craters of active Volcanos, present gradations of structure, and composition, which connect them with the most ancient Porphyries, Sienites, and Granites.

The simplest cases of volcanic action are those of Trachyte (g. 1.) and of Lava (i. 5.) ejected through apertures in Granite; such cases prove that the source of volcanic fires, is wholly unconnected with the pseudo-volcanic results of the combustion of coal, bitumen, or sulphur, in stratified formations, and is seated deep beneath the Primary rocks.†

* The appellation of Trachyte has been given to a volcanic Porphyry, usually containing Crystals of glassy felspar, and remarkably harsh to the touch, (hence its name from τραχὺς); it does not occur in Britain, but abounds in the neighbourhood of almost all extinct and active volcanic craters.

† The occurrence of angular fragments of altered Granite, embedded in Pillars of columnar Lava, in the valley of Monpezat in the Ardêche, shews that these fragments were probably torn off during

Craters.

Our section represents three cases of Volcanic craters; the most simple (i. 5.) rising through Granite, or stratified rocks, at the bottom of the sea, and accumulating craters, which, like those of Lipari and Stromboli, Sabrina, and Graham Islands, are occasionally formed in various parts of the ocean.* The second case is that of volcanos, which, like Etna and Vesuvius, are still in action on the dry land, (i. 1. to i. 4.) The third is that of extinct volcanos, like those in Auvergne, (h¹. h².) which, although there exist no historical records as to the period of their last eruptions, shew by the perfect condition of their craters, that they have been formed since the latest of those aqueous inundations, that have affected the Basalts and Tertiary strata, through which they have burst forth.

One great difference between the more ancient Basaltic eruptions and those of the Lava and Trachyte of existing volcanos, is that the emission of the former, probably taking place under the pressure of deep water, was not accompanied by the formation of any permanent craters.

In both cases, the fissures through some of which these Eruptions may have issued, are abundantly apparent under

the upward passage of the Lava through fractures in the solid Granite.

At Graveneire, near Clermont, a stream of Lava still retains the exact form, in which it issued through a fissure in the side of a mountain of Granite, and overflowed the subjacent valley. Most accurate representations of this, and many similar productions of Volcanic Eruptions from the Granite of this District may be seen in Mr. Poulett Scrope's inimitable Panoramic Views of the Volcanic formations of Central France.

* Within the last few years, the Volcanic Cones of Sabrina in the Atlantic, and of Graham Island in the Mediterranean, have risen suddenly in the sea and been soon levelled and dispersed by the Waves.

the form of Dykes, filled with materials similar to those
which form the masses that have overflowed in the Vicinity of each Dyke.*

*Changes effected by the Igneous Rocks, on the Strata
in contact with them.*

The peculiar condition of the rocks that form the side
walls of Granitic Veins and Basaltic Dykes, affords another argument in favour of their igneous origin; thus
wherever the early Slate rocks are intersected by Granitic
Veins (a. 8.) they are usually altered to a state approxima-
ting to that of fine-grained Mica-Slate, or Hornblende-
Slate.

The Secondary and Tertiary rocks also, when they are
intersected by basaltic Dykes, have frequently undergone
some change; beds of Shale and Sandstone are indurated,
and reduced to Jasper; compact Limestone and Chalk are
converted to crystalline Marble, and Chalk-flints altered to
a state like that resulting from heat in an artificial fur-
nace.†

In all these cases, the Phenomena appear to be throughout consistent with the theory of igneous Injection, and to
be incapable of explanation on any other Hypothesis that
has been proposed. A summary statement of the probable
relations of the Granitic and Trappean Rocks to the other
materials of the Globe, and to one another, may be found
in De la Beche's Geological Researches, 1st Edit. Pag.
374, et seq.

---

* In many dykes the materials have been variously modified, by
their mode of cooling, and differ from the masses which overflowed
the surface.

† Examples of this kind occur on the sides of Basaltic Dykes intersecting Chalk in the County of Antrim, and in the Island of
Raghlin. See Geol. Trans. London, O. S. vol. iii. p. 210. pl. 16.

Explanation of Letters and Figures used in the references to unstratified and crystalline Rocks in Plate 1.

a. Granite.   b. Sienite.   c. Porphyry.

d. Greenstone.   e. Serpentine.   f. Basalt, or Trap.

g. Trachyte.   h. Products of Extinct Volcanos.

i. Products of Active Volcanos.

a. 1.—a. 3. Mountains of Granite, raised into lofty ridges, from beneath Gneiss and Primary Slates.

a. 4. Granite intermixed with Gneiss.

a. 5.—a. 8. Granite, subjacent to stratified rocks of all ages, and intersected by volcanic rocks.

a. 9. Granite Veins, intersecting Granite, Gneiss, and primary Slate.

a. 10. Granite Vein, intersecting Primary and Transition rocks, and forming overlying masses at the surface.

a. 11. Granite Vein intersecting Secondary strata, and overlying Chalk.*

b. Dykes of Sienite.

b. 1. Overlying masses of Sienite.

c. Dykes of Porphyry.

c. 1. Overlying masses of Porphyry.

d. Dykes of ancient Greenstone.

d. 1. Overlying masses of the same. The Rocks represented by d. and e. often pass into one another.

e. Dykes of Serpentine.

e. 1. Overlying masses of Serpentine.

f. Dykes and intruded subterraneous masses of Basalt.

f. 1. to f. 7. Masses of Basalt protruded through, and overlying strata of various ages.

* In the locality quoted in the *Explanation of Plates*, Vol. II. p. 5, the Granite which comes to the surface over the Chalk, is not covered by Tertiary deposits, as represented in our section, Pl. 1.

f. 8. Basaltiform products of Modern Volcanos.

g. Trachyte forming Dykes.

g. 1. Trachyte forming overlying Domes. (Puy-de Dome.)

h. 1. h. 2. Lava of extinct Volcanos, forming undisturbed
  Cones. (Auvergne.)

i.—i. 5. Lava, Scoriæ, and Craters of active Volcanos.
  (i. 1.—i. 4. Etna. 1—5. Stromboli.)

k.—k. 24. Metalliferous Veins.

k. 15'. Lateral expansions of Veins into metalliferous
  cavities, called by the Miners Pipe Veins, or Flats.

l.—l. 7. Faults, or fractures and dislocations of the strata.
  The continuity of stratified Rocks is always inter-
  rupted, and their level more or less changed on the
  opposite sides of a fault.

It is unnecessary here to give detailed descriptions of the
28 divisions of the Stratified Rocks represented in our Section. Their usual Order of Succession and Names are expressed in their respective places, and detailed descriptions
of their several characters may be found in all good Trea-
tises on Geology.

The leading Groups of Formations are united by colours,
marking their separation from the adjacent groups; and the
same colours are repeated, in the headings above the figures
of Plants and Animals, that characterize the several series
of Formations, to shew the extent of the strata over which
the Organic Remains of each Group are respectively distributed.

The Formations of Peat Bogs and Calcareous Tufa are
of too local a nature to be included in the series of *stratified
Rocks* represented in this Section; although they sometimes operate locally to a considerable extent, in adding
permanent and solid matter to the surface of the Globe.

*List of the Names of the Plants and Animals, represented
in Pl. 1. to denote the prevailing Types of Vegetable and
Animal Life, during the formation of the three great
divisions of stratified Rocks.*

REFERENCES.

r. *recent.*   f. *fossil.*   Ad. B. *Adolphe Brongniart.*   L. *Lindley.*
Ag. *Agassiz.*   P. *Page of Vol. I.*

Remains in Transition Strata.

LAND PLANTS.

1. Araucaria.   Norfolk Island Pine.   r. & f.   P. 484.
2. Equisetum.   r. & f.   P. 460.
3. Calamites nodosus.   f.   (L. Pl. 16.)
4. Asterophyllites comosa.   f.   (L. 108.)
5. Asterophyllites foliosa.   f.   (L. 25.)
6. Aspidium.   r.   Pecopteris.   f.
7. Cyathea glauca, Tree Fern.   r.   (Ad. B. Hist. Veg.
   Foss. Pl. 38.)   P. 464.
8. Osmunda.   r.*   Neuropteris.   f.
9. Lycopodium cernuum.   r. (from Mirbel.)   P. 466.
10. Lycopodium alopecuroides.   r.   (from Mirbel.)   P.
    466.
11. Lepidodendron Sternbergii.   f.
12. Lepidodendron gracile?   f.
13. Flabelliform Palm.   r. (from Mirbel.)   Palmacites.   f.

MARINE ANIMALS AND PLANTS.

14. Acanthodes.   f.   Ag.
15. Catopterus.   f.   Ag.
16. Amblypterus.   f.   Ag.
17. Orodus, extinct Shark.   f.   (imaginary restoration).

* An error in this figure represents the fructification as branching
from the tallest frond, instead of rising by a separate rachis from the
root.

18. Cestracion Phillippi, Port Jackson Shark. r. (Phillip.) P. 288.*
18'. Palatal Tooth of Cestracion Phillippi. r.
19. Tooth of Psammodus, from Derbyshire limestone. f.
19'. Tooth of Orodus, from Mountain limestone, near Bristol. f.
20. Calymene. f. ⎫
21. Paradoxus f. ⎬ Trilobites. P. 391.
22. Asaphus. f. ⎭
23. Euomphalus. f.
24. Producta. f.
25. Spirifer. f.
26. Actinocrinites. f. (Miller, P. 96.) P. 417.
27. Platycrinites. f. (Miller, P. 74.) †
27*. Fucoides circinatus. f. (Ad. B.) From Transition sandstone, Sweden.
28. Caryophyllia. r. & f.
29. Astrea. r. & f.
30. Turbinolia. r. & f.

Remains in Secondary Strata.

LAND PLANTS.
31. Pinus. r. & f.
32. Thuia. r. & f.
33. Cycas circinalis. r. Cycadites. f.
34. Cycas revoluta. r. Cycadites. f.
35. Zamia horrida. r. Zamia. f.
36. Dracæna. r. Allied to Bucklandia and Clathraria. f.
37. Arborescent Fern. r. P. 465.
38. Pteris aquilina. r. Pecopteris. f.

* This shark is the only known living representative of the extinct genus Psammodus.

† Fig. 27. In most, if not all the species of Platycrinites the arms are subdivided ; they are not so in this figure, as from its small size they could not well be represented. The figure is intended to give only a general idea of the subject.

39. Scolopendrium. r. Tæniopteris in Oolite. Scarborough. f.

LAND ANIMALS,

40. Didelphys. r. Stonesfield slate, 2 small species. f.
41. Didelphys. r. Cheirotherium? f. P. 265.
42. Pterodactylus brevirostris. f.
43. Pterodactylus crassirostris. f.
44. Gavial. r. Allied to Teleosaurus. f.
45. Iguana. r. Iguanodon. f.
46. Testudo, Land Tortoise. r. Scales of Tortoises, at Stonesfield, Oxon. f. Footsteps of Tortoises, Dumfries. f.
47. Emys. r. Soleure. f.
48. Buprestis. r. Stonesfield. f.
49. Libellula. r. Solenhofen. f.

MARINE ANIMALS, AND PLANTS.

50. Plesiosaurus. f.
51. Ichthyosaurus. f.
52. Marine Turtle. r. At Luneville, in Muschel Kalk. f. P. 256.
53. Pygopterus. f. (Ag. Vol. I. Pl. D. 3.) In Magnesian Limestone.
54. Dapedium, in Lias. f.
55. Hybodus. f. Extinct genus of Sharks. (Imaginary restoration.)
56. Loligo. r. Lyme Regis. f.
57. Nautilus Pompilius. r. Many species. f.
58. Ammonites Bucklandi. f. Peculiar to Lias.
59. Astacus. r. & f.
60. Limulus, King Crab. r. Solenhofen. f.
61. Trigonia. f. New Holland. r.
62. Ophiura. r. & f.
63. Asterias. r. & f.

64. Echinus. r. & f.
65. Apiocrinites. f.
65ᵃ. Fucoides recurvus. f. (Ad. B. Hist. Veg. Foss.
   Pl. 5. Fig. 2.)

Remains in Tertiary Strata.

LAND PLANTS.

66. Mauritia aculeata. r. (Martius, T. 44.) Palmacites.
   Lamanonis. f. P. 214.
67. Elaeis guineensis. r. (Martius, T. 56.) Fruits of
   Pinnate Palms. f. P. 515.
68. Cocos nucifera. r. (Martius, Pl. 62.) Fossil Cocoa
   nut, Sheppy, Brussels. P. 515.
69. Pinus, Pine. r. & f.
70. Ulmus, Elm. r. & f.
71. Populus, Poplar. r. & f.
72. Salix, Willow. r. & f.

LAND ANIMALS OF FIRST PERIOD.

Birds.

73. Scolopax, Woodcock. r. & f.
74. Ibis. r. & f.
75. Tringa, Sea Lark, r. & f.
76. Coturnix, Quail. r. & f.
77. Strix, Owl. r. & f.
78. Buteo, Buzzard. r. & f.
79. Phalacrocorax, Cormorant. r. Pelecanus. f.

Reptiles.

80. Emys, Fresh water Tortoise. r. & f.
81. Trionyx, Soft Tortoise. r. & f.
82. Crocodilus, Crocodile. r. & f.

Mammifers.

83. Vespertilio, Bat. r. & f.
84. Sciurus, Squirrel. r. & f.
85. Myoxus, Dormouse. r. & f.

86. Castor, Beaver.   r. & f.
87. Genetta, Genet.   r. & f.
88. Nasua, Coati.   r. & f.
89. Procyon, Racoon.   r. & f.
90. Canis Vulpes, Fox.   r. & f.
91. Canis Lupus, Wolf.   r. & f.
92. Didelphys, Opossum, small.   r. & f.
93. Anoplotherium commune.   f.
94. Anoplotherium gracile.   f.
95. Palæotherium magnum.   f.
96. Palæotherium minus.   f.

MARINE ANIMALS.

Mollusks.

Genera of Shells most characteristic of the Tertiary Periods ..

    a. Planorbis.   r. & f.
    b. Limnæa.   r. & f.
    c. Conus.   r. & f.
    d. Bulla.   r. & f.
    e. Cypræa.   r. & f.
    f. Ampullaria.   r. & f.
    g. Scalaria.   r. & f.
    h. Cerithium.   r. & f.
    i. Cassis.   r. & f.
    j. Pyrula.   r. & f.
    k. Fusus.   r. & f.
    l. Voluta.   r. & f.
    m. Buccinum.   r. & f.
    n. Rostellaria.   r. & f.

Mammifers.

97. Phoca, Seal.   r. & f.
98. Trichechus, Walrus.   r. & f.
99. Delphinus Orca, (Phocœna, Cuv.) Grampus.   r.   Delphinus.   f.
100. Manatus, Lamantin.   r. & f.
101. Balæna, Whale.   r. & f.

LAND ANIMALS.*

Birds.

Aves........ {
102. Columba, Pigeon. r. & f.
103. Alauda, Lark. r. & f.
104. Corvus, Raven. r. & f.
105. Anas, Duck. r. & f.

Mammifers.

Ruminantia.. {
106. Alces, Elk. r. & f.
107. Elaphus, Stag. r. & f.
108. Bos Urus, Bison. r. & f.
109. Bos Taurus, Ox. r. & f.

Rodentia.... 110. Lepus, Hare. r. & f.

Carnivora... {
111. Ursus, Bear. r. & f.
112. Mustela, Weasel. r. & f.
113. Hyæna. r. & f.
114. Felis, Tiger. r. & f.

Pachydermata {
115. Sus, Hog. r. & f.
116. Equus, Horse. r. & f.
117. Rhinoceros. r. & f.
118. Hippopotamus. r. & f.
119. Elephas. r. Mammoth. f.

Animal of the present Epoch, supposed to have recently become extinct.

120. Didus, Dodo. r. & f.

The bones of the Dodo have been found under lava of unknown age in the Isle of France, and in a cavern in the Island of Roderigue. See Zoological Journal, 1828, p. 554. Loudon's Mag. Nat. Hist. Vol. II. p. 442. and London and Edin. Phil. Mag. Dec. 1832.

* Many of the following genera occur both in the second, third and fourth formations of the Tertiary series, and also in Caverns, Fissures, and Diluvium.

PLATE 2.   V. 1. p. 72.

A. Jaw of Didelphys Bucklandi, (magnified to twice
    nat. size,) in the Collection of W. I. Broderip, Esq.
    and described by him in the Zoological Journal,
    V. III. p. 408, Pl. XI.   (Broderip.)

2. Second molar tooth magnified.

5. Fifth molar tooth still further magnified.

B. Fragment of lower Jaw of a small Didelphys from
    Stonesfield, in the Oxford Museum, (magnified one
    third.) This Jaw has been examined by Cuvier,
    and is figured by M. Prevost, Ann. de Sci. Nat. Avr.
    1825, p. 389, Pl. 18.   The removal of a part of the
    bone displays the double roots of the teeth, in their
    alveoli, and the form of the teeth shews the ani-
    mal to have been insectivorous.   (Original.)

4. Fourth molar tooth magnified.

9. Ninth molar tooth magnified.

C. 1. Lower Jaw of Dinotherium giganteum, (Tapirus
    Giganteus, Cuv.) The length of this Jaw, including
    the Tusk, is nearly four feet. V. I. p. 136. (Kaup.)

2. Lower Jaw and part of upper Jaw of Dinotherium
    medium. (Kaup.)

3. Jaw of Dinotherium medium, exhibiting the Crown
    of five molar teeth, most nearly resembling those of
    a Tapir. (Kaup.) *

---

* All these unique remains of Dinotherium are preserved in the
Museum at Darmstadt; they were found in a Sand pit containing
marine shells at Epplesheim near Alzey, about forty miles N. W. of
Darmstadt, and are described by Professor Kaup.

Bones of Dinotherium have lately been found in Tertiary *Fresh-
water* limestone, near Orthes, at the foot of the Pyrenees; and with
them, remains of a new Genus, allied to Rhinoceros; of several unknown species of Deer; and of a Dog, or Wolf, the size of a Lion.

Our figures of Dinotherium are copied from the Atlas of Kaup's
Description d'Ossemens fossiles de Mammifères, Darmstadt, 1832-3.

PLATE 3.   V. I. p 81.

Imaginary restoration of four species of Pachydermata, found in the Gypsum Quarries of Mont Martre. (Cuvier.)

PLATE 4.   V. I. pp. 81, 85.

Nearly perfect skeletons of the four species of fossil animals, whose restored figures are given in the last Plate. (Cuvier.)

PLATE 5.   V. I. p. 139.

1. Skeleton of Megatherium, copied from Pander and D'Alton's figure of the nearly perfect skeleton of this animal, in the Museum at Madrid.
2. Bones of the Pelvis of Megatherium, discovered by Woodbine Parish, Esq. near Buenos Ayres, and now placed in the Museum of the Royal College of Surgeons, London. The bones of the left hind leg, and several of those of the foot, are restored nearly to their natural place.   (Original.)
3. Front view of the left Femur.
4. Front view of the left Tibia and Fibula.
5. Bones of the foot, imperfectly restored.
5'. Large ungual bone, supposed to be that of a Toe of the hind foot.
6—11. Teeth of Megatherium.

From the near approximation of this Animal to the living Tapir, we may infer that it was furnished with a Proboscis, by means of which it conveyed to its mouth the Vegetables it raked from the bottom of Lakes and Rivers by its Tusks and Claws. The bifid ungual bone (Kaup, Add. Tab. 11,) discovered with the other remains of Dinotherium, having the remarkable bifurcation which is found in no living Quadrupeds, except the Pangolins, seems to have borne a Claw, like that of these animals, possessing peculiar advantages for the purpose of scraping and digging; and indicating functions, concurrent with those of the Tusks and Scapulæ. (see Vol. I. Page 136.)

12, 13. Armour, supposed to be that of Megatherium.*

14—19. Armour of Dasypus and Chlamyphorus.

PLATE 6.   V. I. p. 148.

1. Sections of Teeth of Megatherium, illustrating the relative dispositions of the Ivory, Enamel, and Crusta petrosa, or Cœmentum. (Original. Clift.)

2. Posterior surface of a caudal vertebra of Megatherium, exhibiting enormous transverse processes. . On its lower margin are seen the articulating surfaces which received thé chevron bone; the superior spinous process is broken off. V. I. p: 151. (Sir F. Chantrey. Original.)

PLATE 7.   V. I. p. 168.

Ichthyosaurus platyodon from the Lias at Lyme Regis, discovered by T. Hawkins, Esq. and deposited in the British Museum, together with all the other splendid fossil remains that are engraved in his memoirs of Ichthyosauri and Plesiosauri. This animal, though by no means full grown, must have measured twenty-four feet in length. The extremity of the tail, and left fore paddle, and some lost

---

* Mr. Darwin has recently discovered the Remains of Megatherium along an extent of nearly six hundred miles, in a North and South line, in the great sandy plains of the Pampas of Buenos Ayres, accompanied by bones and Teeth of at least five other Quadrupeds. He has also found that the Bones of this Animal are so often accompanied by those of the Mastodon angustidens, as to leave no doubt that these two extinct species were contemporary.

I learn from Professor Lichstenstein, that a fresh importation of Bones of Megatherium, and bony armour has lately been sent to Berlin from Buenos Ayres, and that there remains no room to doubt that some portion of this armour appertained to the Megatherium.

It appears very probable, from more recent discoveries, that several other large and small animals, armed with a similar coat of mail, were co-inhabitants of the same sandy regions with the Megatherium.

fragments of the rest of the skeleton, are artificially restored. (Hawkins.)

PLATE 8.   V. I. p. 170.

1. Skeleton of a young Ichthyosaurus communis, in the collection of the Geological Society of London, found in the Lias at Lyme Regis. (Scharf. Original.)
2. Ichthyosaurus intermedius, from Lyme Regis, belonging to Sir Astley-Cooper. (Scharf. Original.)

PLATE 9.   V. I. p. 170.

1 and 2. Ichthyosaurus tenuirostris, from the Lias near Glastonbury, in the collection of the Rev. D. Williams, of Bleadon, near Bristol. The position of the ribs is distorted by pressure. (Scharf. Original.)
3. View of the right side of the head of the same animal. (Original.)

PLATE 10.   V. I. pp. 171, 173.

1. Head of Ichthyosaurus platyodon, in the British Museum, from the Lias at Lyme Regis, copied from Sir E. Home's figure in the Phil. Trans. 1814.
2. Copied from Mr. Conybeare's figure, (in the Geol. Trans. Lond. O. S. Pl. XL. Fig 11.) shewing the analogies between the bones of the head of Ichthyosaurus, and those which Cuvier has marked by corresponding letters in his figure of the head of the Crocodile.
3. Two of the bony plates in the sclerotic coat of the Eye of Ichthyosaurus platyodon.
4. Circle of bony plates in the Eye of the snowy Owl. (Yarrel.)
5. Circle of similar plates in the Eye of the golden Eagle. (Yarrel.)

6. Front view of bony plates in the Eye of an Iguana.

7. Profile of the same.

8. Two of the fourteen component scales of the same.

I owe these three last figures to the kindness of Mr. Allis of York.

A 1, 2, 3, 4. Petrified portions of the skin of a small Ichthyosaurus, from the Lias of Barrow on Soar, Leicestershire, presented to the Oxford Museum, by the Rev. Robert Gutch, of Segrave. (Original.)

In Fig. 1; a, b, c, d, are portions of ribs, and e, f, g, h, are fragments of sterno-costal bones (nat. size).

The spaces between these bones, are covered with the remains of skin; the Epidermis being represented by a delicate film, and the Rete mucosum by fine threads of white Carbonate of Lime; beneath these the Corium, or true skin, is preserved in the state of dark Carbonate of Lime, charged with black volatile matter, of a bituminous and oily consistence.

2. Magnified representation of the Epidermis and Rete mucosum. The fine superficial lines represent the minute wrinkles of the Epidermis, and the subjacent larger decussating lines, the vascular net-work of the Rete mucosum.

In Fig. 3, the Epidermis exhibits a succession of coarser and more distant folds or wrinkles, overlying the mesh-work of the Rete mucosum.

In Fig. 4, the Epidermis has perished, and the texture of the fine vessels of the Rete mucosum is exhibited in strong relief, over the black substance of the subjacent Corium, in the form of a net-work of white threads.*

---

* Nothing certain has hitherto been known respecting the dermal covering of the Ichthyosauri; it might have been conjectured that these reptiles were incased with horny scales, like Lizards, or that their skin was set with dermal bones, like those on the back of Cro-

PLATE 11.   V. I. p. 175.

1. Side View of the head of an Ichthyosaurus, marking
   by corresponding letters, the analogies to Cuvier's
   figures of the same bones in the head of the Crocodile.   (Conybeare.)
2. Posterior part of a lower jaw of Ichthyosaurus communis, in the Oxford Museum.   (Conybeare.)
3—7. Sections presented by the component bones of Fig.
   2 in fractured parts above each section. (Conybeare.)
8. View of the lower Jaw of Ichthyosaurus seen from

codiles; but as the horny scales of Fishes, and dermal bones of Crocodilean animals are preserved in the same Lias with the bones of
Ichthyosauri, we may infer that if the latter animals had been furnished with any similar appendages, these would also have been preserved, and long ere this discovered, among the numerous remains
that have been so assiduously collected from the Lias. They would
certainly have been found in the case of the individual now before
us, in which even the Epidermis and vessels of the Rete Mucosum
have escaped destruction.

Similar black patches of petrified skin are not unfrequently found
attached to the skeletons of Ichthyosauri from Lyme Regis, but no
remains of any other soft parts of the body have yet been noticed.

The preservation of the skin shews that a short interval only elapsed
between the death of the animal, and its interment in the muddy se-
diment of which the Lias is composed.

Among living reptiles, the Batrachians afford an example of an
order in which the skin is naked, having neither scales nor dermal
bones.

In the case of Lizards and Crocodiles, the scaly, or bony coverings
protect the skin from injury by friction against the hard substances
with which they are liable to come into contact upon the land; but
to the Ichthyosauri which lived exclusively in the sea, there would
seem to have been no more need of the protection of scales or
dermal bones, than to the naked skin of Cetacea.

In the case of Plesiosauri also, the non-discovery of the remains
of any dermal appendages with the perfect skeletons of animals of
that genus, leads to a similar inference, that they too had a naked
skin. The same negative argument applies to the flying Reptile
Family of Pterodactyles.

beneath, exhibiting the course of its over lapping
bones. (Conybeare.)

A. Tooth of a Crocodile, shewing the incipient absorption of the hollow cone which forms its base, from
the effect of pressure of a new tooth rising beneath.
(Conybeare.)

B. Similar effects shewn in the transverse section of the
upper and lower jaws of an Ichthyosaurus. (Cuvier.)

C. Example of the same kind of absorption produced by
the pressure of a new tooth, on the base of an older
tooth in the jaw of Ichthyosaurus. (Conybeare.)

PLATE 12.   V. 1. p. 181.

1. Sternal Arch and Paddles of Ichthyosaurus. See V. I.
   p. 182, Note. (Home.)

2. Sternal Arch of Ornithorhynchus. (Home.)

3, 4, 5, 6. Occipital and Cervical Bones of Ichthyosaurus, from the Lias at Lyme Regis.* (Original.)

* Sir Philip de Malpas Grey Egerton has pointed out some beautiful examples, hitherto unnoticed, in the Atlas and cervical Vertebræ of Ichthyosauri, of peculiar mechanical contrivances to support and regulate the movements of their enormous heads. (See
Lond. and Edin. Phil. Mag. Nov. 1835. p. 414.)

Fig. 3, a. represents the Basilar portion of the Occipital bone of a
very large and aged Ichthyosaurus, from the Lias of Lyme Regis,
(scale one eighth). The nearly hemispherical process (a) articulated
with a comparatively shallow socket in front of the Atlas, (4. a.)
and this ball and socket, or universal joint, gave freedom of motion
and support to a weighty head.

Fig. 4. Atlas and Axis of a very young Ichthyosaurus, (two thirds
of nat. size.) These bones adhere together by two nearly flat surfaces, admitting of the least flexure of any of the Vertebræ in the
whole body, but giving the greatest strength to that part of the
Column, where strength rather than flexure was required.

On the inferior margins of the Atlas and Axis and third cervical
vertebra, are triangular facets articulating with three strong wedge-
shaped sub-vertebral bones (c) hitherto undescribed.

A. Hollow conical Vertebræ of a fish. (Original.)

B. C. E. Vertebræ of Ichthyosaurus. See Note, V. I.
    p. 178. (Home and Conybeare.)

D. a. g. E. a. g. Spinous processes, shewing the peculiar
    articulation of their annular portions, with the Ver-
    tebræ, to be adapted to increase the flexibility of the
    spine. See Note, V. I. p. 179. (Home.)

PLATE 13.   V. I. p. 190.

Skeleton of a small Ichthyosaurus, from the Lias at

Fig. 4, b. Oblique triangular facet on the lower margin of the
front of the Atlas; this facet articulated with the first sub-vertebral
wedge, placed between the Atlas and Occiput.

Between the Atlas and Axis, the two sub-vertebral facets formed a
triangular cavity for the reception of a second wedge (Fig. 4. c.)
and a similar, but smaller cavity received another wedge of the
same kind, between the Axis and third Vertebra. This third wedge
gave less support to the head, and admitted of more extensive motion
than the second. All these three wedge-shaped bones are seen
nearly in their natural position in a specimen from Lyme Regis, in
the Collection of Sir P. G. Egerton.

Fig. 4′. First sub-vertebral wedge, auxiliary to the anterior cavity
of the Atlas, in completing the articulating socket for the basilar
process of the Occiput (3. a).

4. a. Crescent shaped front of the first sub-vertebral wedge.

4′. b. Head of the same Wedge.

4′. c. Obtuse apex of the same, articulating with the triangular
frontal facet of the Atlas (4. b.). In young animals this frontal facet
is nearly smooth and flat; in older animals (3. b′.) it is rugged and
furrowed. This articulation must have given to the first sub-vertebral
wedge great power as a stay or prop, to resist the downward pressure of the head, at the same time facilitating the rotatory movements of the Occipital bone.

Fig. 4. c. Second sub-vertebral wedge articulating with the triangular cavity formed by the marginal facets of the Atlas and Axis.
This second Wedge acted as a strong prop supporting firmly the
lower portion of the Atlas, and at the same time admitting the small
amount of motion here required.

c′. Head of the sub-vertebral wedge (c) strengthened by a projecting boss of solid bone.

Lyme Regis, presented to the Oxford Museum by Viscount Cole, enclosing within its ribs scales, and digested bones of Fishes, in the state of Coprolite. This coprolitic mass seems nearly to retain the form of the stomach of the animal. c, Coracoid bone.   d, Scapula.   e, Humerus.   f, Radius. g, Ulna.   (Scharf.   Original.)

PLATE 14.   V. I. p. 191.

Skeleton of the Trunk of a small Ichthyosaurus in the

Fig. 5. Nearly flat articular surface of (probably) the third cervical vertebra of the same large Individual as Fig. 3. This surface of the bone has only a small cylindrical depression at its centre, instead of the deep, conical cup of the more flexible vertebræ, C. B. E.

· Near its upper margin is a wedge-shaped elevation (b) and near the inferior margin, a notch or furrow (a). These salient and re-entering portions articulated with corresponding depressions and projections on the surface of the adjacent vertebra, and acted as pivots, admitting a limited amount of lateral vibrations, and at the same time preventing any slip, or dislocation.

Fig. 6. Concave surface of Fig. 5.; the wedge-shaped projection near its lower margin (a) must have articulated with a corresponding groove or depression on the front of the vertebra adjacent to it, like that at (Fig. 5. a.) As one surface only of these vertebræ had a conical cavity, the intervertebral substance must have formed a single cone, admitting in the neck but half the amount of motion, that the double cones of intervertebral matter allowed to the dorsal and caudal vertebræ, (C. B. E.) where greater flexure was required, to effect progressive motion by vibrations of the body and tail.

These dispositions of the articulating facets of the cervical vertebræ, acting in conjunction with the three sub-vertebral wedges before described, afford an example of peculiar provisions in the neck of these gigantic Reptiles, to combine a diminished amount of flexure in this part, with an increased support to their enormous heads.

It is probable that every species of Ichthyosaurus had peculiar variations in the details of the cervical vertebræ, and subvertebral wedges, and that in each species these variations were modified by age.

In the Gavial Mr. Mantell has recently observed that the first caudal vertebra is *doubly convex*, like the last cervical vertebra in Turtles. These peculiar contrivances give to the animals in which they occur increased flexibility of the Tail and Neck.

Oxford Museum, from the Lias at Lyme Regis, containing within the ribs, a coprolitic mass of digested bones, interspersed with scales of fishes. a, Furcula. b, Clavicle. c, Coracoid bone. d, Scapula. e, Humerus. (Fisher. Original.)

PLATE 15.   V. I. p. 187.)

The specimens are all of the natural size except where the figures denote otherwise. (Original.)

1 and 2. Intestines of the two most common English species of Dog-Fishes, injected with Roman cement.
The vascular structure, which is still apparent in the desiccated membrane, resembles the impressions on the surface of many Coprolites.

3. Coprolite from the Lias at Lyme, exhibiting the spiral folding of the plate of digested bone, and impressions of the intestinal vessels and folds upon its surface. (See Note, V. I. p. 194. et seq.)

3'. Magnified scale of Pholidophorus limbatus, embedded in the surface of the Coprolite, Fig. 3. This scale is one of those that compose the lateral line, by which a tube passes to convey mucus, from the head, along the body of fishes; a. is the hook, on the superior margin, which is received by a depression on the inferior margin of the scale above it, corresponding with b.; c. is the serrated edge of the posterior margin, perforated at e. for the passage of the mucous duct; d. is a tube on the interior surface of the scale to carry and protect the mucous duct. (See note V. I. p. 191.)

3''. Exterior of the scale 3'.; the same parts are represented by the same letters; the larger portion is covered with enamel; the smaller portion next d. is the bony root forming the anterior margin of the scale.

4. Transverse section of another Coprolite from Lyme,
   shewing the internal foldings of the plate, with sections of scales of fishes embedded in it.

5. Exterior of a spiral Coprolite, from the Chalk Marl,
   near Lewes, shewing folds and vascular impressions
   analogous to those in No. 3.

6. Longitudinal section of another Coprolite, from the
   same Chalk Marl, shewing the spiral manner in
   which the plate was folded round itself.

7. Exterior of another spiral Coprolite, from the Chalk
   at Lewes, shewing vascular impressions on its surface, and the transverse fracture of the spiral fold
   at b.

   In many other figures of Plate 15, a similar
   abrupt termination of the coiled plate is visible at b.

8, 9. Two other small species of spiral Coprolites in chalk;
   these as well as Figs. 5, 6, 7, are probably derived
   from fishes found with them in the chalk, near Lewes.

10, 11, 12. Coprolites from the Lias at Lyme, exhibiting
   well defined characters of the spiral fold, with vascular impressions on their surface.

13. Similar appearances on a Coprolite found by Dr.
   Morton in the Greensand of Virginia.

14. Coprolite from the Lias at Lyme, bearing strong corrugations, the result of muscular pressure received
   from the intestines.

15. Transverse section, shewing the abrupt termination
   of the folded plate in Fig. 14, and representing the
   flattened form of the spiral intestine.

16. Longitudinal section of the intestinal tube of a recent
   Shark, shewing the spiral valve that winds round
   its interior, in the form of an Archimedes screw; a
   similar spiral disposition of the interior is found in
   intestines of Dog-Fishes, Figs. 1 and 2.

17. Coprolite from Lyme, containing large scales of Dapedium politum.

18. Coprolite from the Lias at Lyme, containing undigested bones of a small Ichthyosaurus.

PLATE 15'.   V. I. p. 200.

Cololite, or petrified intestines of a fossil fish from Solenhofen.   (Goldfuss.)

PLATE 16.   V. I. p. 202.

1. Conjectural Restoration of the Skeleton of Plesiosaurus dolichodeirus.   (Conybeare.)

2. Skeleton of Plesiosaurus dolichodeirus, in the British Museum, from the Lias at Lyme Regis.   (Scharf. Original.)

PLATE 17.   V. I. p. 203.

A nearly entire and unique skeleton of Plesiosaurus dolichodeirus, 5 feet 7 inches long, from the Lias of Street, near Glastonbury.   This skeleton forms part of the splendid series of fossil Saurians, purchased for the British Museum, from T. Hawkins, Esq. in 1834. See V. I. p. 208, and Note. (Hawkins.)

PLATE 18.   V. I. p. 205, Note.

1. Under jaw of Plesiosaurus dolichodeirus, forming part of the series last mentioned. (Original.)

2. Head of the Plesiosaurus, figured in Pl. 16. Fig. 2. seen from beneath. (Original.)

3. Ventral portion of the ribs of the Plesiosaurus, figured in Pl. 17. See V. I. p. 208. (Original.)

a. c. Central bones forming the crown of the sterno-costal arch. b. triple series of intermediate bones between the central bones, a. c. and the true ribs, d. d. e. e. lower extremity of coracoid bones.

PLATE 19. V. I. p. 205.

Fig. 1. A beautiful specimen of Plesiosaurus macroce-
phalus hitherto undescribed, found in the Lias marl
at Lyme Regis by Miss Anning, and now in the
collection of Lord Cole. (Original.)

On comparing this figure with those of P. Dolichodeirus
at Pl. 16, 17. The following differences are obvious :

1. The head is very much larger and longer, being
nearly one half the length of the neck.

2. The vertebræ of the neck are thicker and stronger in
proportion to the greater weight they had to sustain.

3. The hatchet-shaped bones differ in form and size, as
may be seen by comparing them with those of P. dolichodeirus. Pl. 19. Fig. 2. and Pl. 17.

4. The bones of the arm and thigh are shorter and
stronger than in P. Dolichodeirus, and corresponding differences may be traced throughout the smaller bones of the
Paddles; the general adjustment of all the proportions
being calculated to produce greater strength in the P. Ma-
crocephalus, than in the more slender limbs of P. Dolichodeirus.

These differences are not the effect of age ; as the two
specimens, from which they are here described, are nearly
of the same length.

Fig. 2. Hatchet-shaped bones of the neck of Plesiosaurus
Dolichodeirus, copied from the specimen figured in
Pl. 17.

3. Anterior extremity of an insulated lower Jaw of Plesiosaurus, from the Lias at Lyme Regis, in the British Museum, part of the collection of Mr. Hawkins.
V. I. p. 205. Note. (Original.)

4. The entire bone, of which Fig. 3. forms part, reduced
to a small scale.

PLATE 20.   V. I. p. 215.

Head of the Mosasaurus, or Great Animal of Maestricht.
(Cuvier.)

PLATE 21.   V. I. p. 221.

Pterodactylus longirostris in the lithographic slate of
Aichstedt.   (Cuvier and Goldfuss.)

In this Plate, and Plate 22, the same letters and figures
designate the corresponding Bones in the different Animals
to which they are affixed ; they are copied chiefly from the
figures and Explanations of Dr. Goldfuss, in his Beiträge
zur Kenntniss verschiedener Reptilien der Vorwelt.

Γ. Cavitas narium.

Δ. Cavitas intermedia.

Θ. Orbita.

a. Maxilla superior.

b. Vel os nasi vel inter-maxillare?

c. Operculum nasale.

d. Aut os frontis anterius vel nasale?

e. Os frontis proprium.

f. Os parietale.

g. Os petrosum.

H. Pars basilaris ossis occipitis.

h. Pars lateralis.

i. Os tympanicum s. quadratum.

k. Os frontis posterius.

l. Os mastoideum.

m. Os zygomaticum.

n. Aut os lacrymale vel superciliare?

o. Annulus orbitalis.

P. Corpus ossis sphenoidei.

p. Processus transversus ossis sphenoidei.

q. Os pterygoideum.

r. Os transversum.

s. Os palatinum.

t. Processus palatinus maxillæ superioris.

v. Pars angularis inferior maxillæ inferioris.

w. Pars angularis superior.

x. Pars condyloidea.

y. Pars complementaria, Cuv. (coronalis, auctor.)

z. Os hyoideum.

I. Atlas.

II. Epistropheus.

III—VII. Vertebræ colli.

VIII—XXII. Vertebræ dorsi.

XXIII. XXIV. Vertebræ lumborum.

XXV. XXVI. Os sacrum.

XXVII. Ossa coccygea.

XXVIII. Sternum.

1—15. Costæ.

16. Scapula.

17. Os coracoideum.

18. Ilium.

19. Os pubis.

20. Os ischium.

21. Humerus.

22. Ulna.

23. Radius.

24. Carpus.

25. Os metarcarpi primum s. pollicis.

26. O. m. secundum.

27. O. m. tertium.

28. O. m. quartum.

29. O. m. quintum.

30, 31. Phalanges pollicis.

32—34. Ph. indicis.

35—38. Ph. digiti medii.

39—43. Ph. digiti annularis.

44—47. Ph. digiti auricularis.

48. Femur.

49. Tibia.

50. Fibula.

51. Tarsus.

52—56. Metatarsus.

57, 58. Phalanges digiti primi.

59—61. Ph. d. secundi.

62—65. Ph. d. tertii.

66—70. Ph. d. quarti.

71—74. Ph. d. quinti.

δ Impressions of the membrane of the wing? *

PLATE 22.   V. I. p. 221.

A. Restoration of the Skeleton of Pterodactylus crassirostris. (Goldfuss.)

B. Fore foot of a Lizard. (Cuvier.)

C. Restoration of the right fore foot, or right hand of Pterodactylus crassirostris. (Goldfuss.)

D. The right fore foot, or hand of P. longirostris. (Cuvier and Soemmerring.)

E. The Fore foot of P. macronyx. (Buckland, Geol. Trans. Lond. 2d Ser. Vol. 3. Pl. 27.)

F. The Hind foot of a Lizard. (Cuvier, Oss. Foss. Vol. V. Pt. II. Pl. XVII.)

G. Right foot of P. crassirostris, as conjecturally re-

---

* Professor Agassiz considers that the Corrugations on the surface of the Stone (δ), which Dr. Goldfuss supposed to be the impressions of Hairs, or Feathers, are only casts of the minute foldings of the contracted membrane of the wing. It is probable that Pterodactyles had a naked skin, like the Ichthyosaurus; (See Pl. 10. A.) because the weight of scales would have encumbered their movements in the air.

stored by Dr. Goldfuss. No authority for this seems
to be afforded by the fossil specimen N.

H. Right foot P. longirostris. (Cuvier.)

I. Foot of P. macronyx. (Buckland.)

K. Hind foot of a Bat.

L. Skeleton of Draco volans. (Carus. Comp. Anat. P.
370.) shewing the elongated bones, or false ribs,
which support the membranous expansion of its
Parachute.

M. Skeleton of a Bat. (Cheselden.)

N. Skeleton of P. crassirostris, in the Museum at Bonn,
in Solenhofen slate. (Goldfuss.)

O. Skeleton of P. brevirostris, from near Aichstadt, in
the same slate. (Goldfuss.)

P. Imaginary restoration of Pterodactyles, with a co-
temporary Libellula, and Cycadites.

PLATE 23.   V. I. p. 234.

Fig. 1′. Anterior extremity of the right jaw of Megalosaurus, from the Stonesfield slate, Oxon. (Buckland.)

Fig. 2′. Outside view of the same, exhibiting near the
extremity, large perforations of the bone for the
passage of vessels. (Buckland.)

Fig. 1. Tooth of Megalosaurus, incomplete towards the
root, and seen laterally as in Fig. 1′. Nat. size.
(Buckland.)

Fig. 2. Side view of a tooth nearly arrived at maturity.
The dotted lines mark the compressed conical cavity,
containing Pulp, within the Root of the growing
tooth. Scale two thirds. (Buckland.)

Fig. 3. Transverse section of Fig. 1′. shewing the thick-
ness of the largest tooth (a.) and its root set deep
and firmly in the bony socket, which descends

nearly to the bottom of the Jaw.   Scale two thirds.
(Buckland.)

Fig. 4. Transverse section of the tooth (Fig. 2.) shewing
the manner in which the back and sides are enlarged, and rounded in order to give strength, and
the front brought to a strong and thin cutting edge
at D'. (Buckland.)

PLATE 24.   V. I. p. 240.

Fossil Teeth and bony nasal horn of Iguanodon; and
lower Jaw and Teeth of Iguana. (Mantell and Original.)

In Mr. Mantell's collection there is a perfect thigh bone
of this animal, 3 feet 8 inches long, and 35 inches in circumference at its largest and lower extremity.

PLATE 25.   V. I. p. 249.

Fig. 1. Fossil Crocodilean found at Saltwick near
Whitby, eighteen feet long, and preserved in the
Museum of that town.   This figure is copied from
Plate XVI. of Bird and Young's Geol. Survey of
the Yorkshire coast.   As this appears to be the
same species with that engraved in the Phil. Trans.
1758, Vol. 50. Pt. 2. Tab. 22. and Tab. 30, and
presented to the Royal Society by Captain Chap-
man, Mr. König has applied to it the name of
Teleosaurus Chapmanni.

Fig. 2. Another head of Teleosaurus Chapmanni, also in
the Museum at Whitby, and from the Lias of that
neighbourhood.   (Original.)

Fig. 3. Head of a third Individual of the same species
from the same locality, placed in 1834, in the British
Museum, showing the outside of the lower Jaw.
(Young and Bird.)

Fig. 4. View of the inside of a lower Jaw of the same

species, in the Oxford Museum, from the Great
Oolite, at Enslow, near Woodstock, Oxon. (Original.)

PLATE 25′. V. I. p. 251.

Fig. 1. Head of a Crocodile found in 1831, by E. Spencer,
Esq. in the London Clay, of the Isle of Sheppy.
See V. I. p. 251. (Original.)

Fig. 2. Extremity of the upper and lower Jaw of Teleosaurus in the Oxford Museum, from the Great
Oolite at Stonesfield, Oxon. See V. I. p. 252.
(Original.)

Fig. 3. Anterior extremity of the upper Jaw of Steneosaurus, in the Museum of Geneva, from Havre;
the same species occurs in the Kimmeridge Clay
of Shotover hill, near Oxford. See V. I. p. 251.
(De la Beche.)

Fig. 4. Fossil Turtle, from the slate of Glaris. See V. I.
p. 257. (Cuvier.)

PLATE 26. V. I. p. 259.

Fossil Footsteps indicating the Tracks of ancient animals, probably Tortoises, on the New Red Sandstone near
Dumfries. (From a cast presented by Rev. Dr. Duncan.)

PLATE 26′. V. I. p. 263.

Fig. 1. Impressions of footsteps of several unknown
animals upon a slab of New Red Sandstone found
at the depth of eighteen feet in a quarry at Hessberg, near Hildburghausen in Saxony. (Sickler.)

The larger footsteps a. b. c. are referred to an
animal named provisionally, Chirotherium. The
fore feet of this animal were less by one half than
the hind feet, and the tracks of all the feet are

in the same straight line. The footsteps d. e. f. form part of another track of the same kind. Some of the large toes of the Chirotherium, and also of the smaller species, have left distinct impressions of nails: g. h. i. k. l. m. n. o. p. q. form the track of an animal of another species, probably a Tortoise, crossing the same slab, in a different direction.

The irregular cylindrical concretions that intersect each other on the surface of this slab, appear to have been formed in cracks, caused by the contraction of a thin bed of green marl, interposed between two deposits of Sandstone. See note, V. I. p. 266.

Fig. 2. One of the impressions of the hind feet of Chirotherium, on the slab Fig. 1 ; half nat. size. (Sickler.)

Fig. 3. One of the footsteps in the track of the smaller animal, upon this slab ; nat. size. (Sickler.)

M. Link has made out the footsteps of four species of animals in the Hildburghausen sandstone; and it has been conjectured that some of these have been derived from gigantic Batrachians.

PLATE 26″. V. I. p. 265.

Impression of the hind foot of Chirotherium, selected from a well preserved Track upon a slab of sandstone from Hildburghausen, in the British Museum. (Original.)

PLATE 26‴. V. I. p. 265.

Footsteps of a small web-footed animal, probably crocodilean, drawn from a Cast of impressions on Sandstone, found near Hildburghausen. (Original.)

The Sandstones in which all these fossil footsteps have been discovered in Germany and Scotland, appear to be referrible to the same division of the secondary strata, which lies in the middle region of that large, and widely extended series of Sandstones, and Conglomerates, Limestones, and Marls, which English Geologists have usually designated by the common appellation of the *New red Sandstone* Group, including all the strata that are interposed between the Coal formation, and the Lias.

M. Brongniart, in his *Terrain de l'Ecorce du Globe*, 1829, has applied to this middle division the very appropriate name of *Terrain Pœcilien*, (from the Greek ποικίλος), a term equivalent to the names Bunter Sandstein, and Grès bigarré, which it bears in Germany and France; and indicating the same strata which, in England, we call the new Red Sandstone. (See Plate 1. Section No. 17.)

Mr. Conybeare, in his Report on Geology to the British Association at Oxford, 1832 (Page 379, and P. 405, Note), has proposed to extend the term *Pœcilitic* to the entire Group of strata between the Coal formation and the Lias; including the five formations designated in our section (Pl. 1, No. 15, 16. 17, 18, 19), by the names of New Red Conglomerate, Magnesian Limestone, Variegated Sandstone, Shell Limestone, and Variegated Marl. Some common appellative for all these formations has been long a desideratum in Geology; but the word *Pœcilitic* is in sound so like to *Pisolite*, that it may be better to adhere more literally to the Greek root ποικίλος, and apply the common name of *Poikilitic* group to the strata in question.*

---

* The general reception of such a common name for all these strata, and the separation of the Grauwacké series into the *Cambrian* and *Silurian* systems, as proposed by Professor Sedgwick and Mr. Murchison, will afford three nearly equal and most convenient groups or systems, into which the strata composing the Transition and Secondary series may respectively be divided; the former comprehending the Cambrian, Silurian, and Carboniferous systems, and the latter comprehending the Poikilitic, Oolitic, and Cretaceous Groups.

PLATE 26ᵃ.

Ornithichnites, or foot-marks of several extinct species of birds, found in the New Red sandstone of the Valley of the Connecticut.*   (Hitchcock.)

* In the American Journal of Science and Arts, January, 1836. V. XXIX. No. 2. Professor Hitchcock has published a most interesting account of his recent discovery of *Ornithichnites*, or footmarks of birds in the New Red sandstone of the valley of the Connecticut. These tracks have been found at various depths beneath the actual surface, in quarries of laminated flag stones, at five places near the banks of this river, within a distance of thirty miles. The sandstone is inclined from 5°, to 30°, and the Tracks appear to have been made on it before the strata received their inclination. Seven of these tracks occur in three or four quarries within the space of a few rods square; they are so distinct, that he considers them to have been made by as many different species, if not genera, of birds. (See Pl. 26ᵃ, Figs. 1—14.)

The footsteps appear in regular succession, on the continuous track of an animal in the act of walking or running, with the right and left foot always in their relative places.

The distance of the intervals between each footstep on the same track is occasionally varied, but to no greater amount than may be explained by the Bird having altered its pace. Many tracks of different individuals and different species are often found crossing one another; they are sometimes crowded like impressions of feet on the muddy shores of a stream, or pond, where Ducks and Geese resort. (See Pl. 26ᵃ. Figs. 12. 13. 14.)

None of the footsteps appear to be those of Web-footed Birds; they most nearly resemble those of Grallæ, (Waders) or birds whose habits resemble those of Grallæ. The impressions of three toes are usually distinct, except in a few instances; that of the fourth or hind toe is mostly wanting, as in the footsteps of modern Grallæ.

The most remarkable among these footsteps, are those of a gigantic bird, twice the size of an Ostrich, whose foot measured fifteen inches in length, exclusive of the largest claw, which measured two inches. All the three toes were broad and thick. (Pl. 26ᵃ. Fig. 1. and Pl. 26ᵇ. Fig. 1.) These largest footsteps have as yet been found in one quarry only, at Mount Tom near Northampton; here, four nearly parallel tracks of this kind were discovered, and in one of them six footsteps appeared in regular succession, at the distance of

The fossil tracks on this Plate are all nearly on the same scale: viz. one-twenty-fourth. The recent footsteps are on a larger scale.

four feet from one another. In others the distance varied from four to six feet; the latter was probably the longest step of this gigantic bird while running.

Next in size to these are the footsteps of another enormous bird (Pl. 26ª. Fig. 4.) having three toes of a more slender character, measuring from fifteen to sixteen inches long, exclusive of a remarkable appendage extending backwards from the heel eight or nine inches, and apparently intended, like a snow shoe, to sustain the weight of a heavy animal walking on a soft bottom. (See Pl. 26ᵇ. Fig. 2.) The impressions of this appendage resemble those of wiry feathers, or coarse bristles, which seem to have sunk into the mud and sand nearly an inch deep; the toes had sunk much deeper, and round their impressions the mud was raised into a ridge several inches high, like that around the track of an Elephant in Clay. The length of the step of this Bird appears to have been sometimes six feet. On the other tracks the steps are shorter, and the smallest impression indicates a foot but one inch long, with a step of from three to five inches. (Pl. 26ª. 2. 3. 5—14.)

In every track the length of the step increases with the size of the foot, and is much longer in proportion than the steps of any existing species of birds; hence it is inferred that these ancient birds had a greater length of leg than even modern Grallæ. The steps at four feet asunder probably indicate a leg of six feet long.

In the African Ostrich, which weighs 100lbs, and is nine feet high, the length of the leg is about four feet, and that of the foot ten inches.

All these tracks appear to have been made on the Margin of shallow water that was subject to changes of level, and in which sediments of sand and mud were alternately deposited, and the length of leg, which must be inferred from the distance of the footsteps from each other, was well adapted for wading in such situations. No Traces of any Bones but those of fishes (Palæothrissum) have yet been found in the rock containing these footsteps, which are of the highest interest to the Palæontologist, as they establish the new fact of the existence of Birds at the early epoch of the New Red sandstone formation; and further shew that some of the most ancient forms of this class attained a size, far exceeding that of the largest among the feathered inhabitants of the present world, and were adapted for wading and running, rather than for flight.

Fig. 1. Ornithichnites giganteus. Many tracks of this species occur at Mount Tom, near Northampton, U. S.

Fig. 2. O. tuberosus. Portions of three tracks, and a single footstep of a fourth appear on the same slab. The two longest of them are in opposite directions.

Fig. 3. O. tuberosus, on a slab in front of the Court House in Northampton, from Mount Tom.

Fig. 4. O. ingens, from a quarry called the Horse Race, near Gill. The appendage to the heel is not distinct in this track.

Fig. 5. O. diversus, on a flag-stone near the first church door at Northampton, U. S.

Fig. 6. O. diversus. We have here three rows of tracks and a single footstep, from the Horse Race Quarry. These tracks shew no marks of any appendage to the heel.

Fig. 7. O. diversus; found near South Hadley, U. S.

Fig. 8. O. diversus; curvilinear track from the Horse Race Quarry.

Fig. 9. O. diversus. Two parallel tracks from the Horse Race Quarry.

Fig. 10. O. diversus; nearly parallel tracks of two birds, with an appendage behind each foot; from the quarries at Montague, U. S.

Fig. 11. O. minimus; common at the Horse Race Quarry; similar impressions of the feet of small birds vary from half an inch to an inch and half in length.

Figs. 12. 13. 14. O. diversus; from the Horse Race Quarry. Tracks of different individuals of different species, and different sizes cross one another confusedly in these three slabs.

Fig. 15. Recent track of probably a Snipe.

Fig. 16. Recent track of a Pea-hen.

Fig. 17. Recent track of a domestic hen.

PLATE 26ᵇ.

Fig. 1. Ornithichnites giganteus. The natural cast here figured represents the form and size of the foot, and part of the claws. (Hitchcock.)

Fig. 2. Ornithichnites diversus; with impressions of the appendage to the heel, drawn from a plaster mould sent by Prof. Hitchcock to the Geol. Soc. of London. (Original.)

Fig. 3. Track of a small animal on Oolitic slate near Bath. See Journal of Royal Institution of London, 1831, p. 538, Pl. 5. (Poulett Scrope.)*

PLATE 27. V. I. p. 269.

Figs. 1—8. Tubercles and Scales, illustrating the four new Orders of Fishes, established by Professor Agassiz. (Agassiz.)

* Mr. Poulett Scrope has presented to the Geol. Soc. of London a series of Slabs selected from the tile quarries worked in the Forest Marble beds of the Oolite formation near Bradford and Bath. The surface of these beds is covered with small undulations or ripple markings, such as are common on the sand of every shallow shore, and also with numerous tracks of small animals (apparently Crustaceans) which traversed the sand in various directions, whilst it was yet soft, and covered with a thin film of clay. These footmarks are in double lines parallel to each other, shewing two indentations, as if formed by small claws, and sometimes traces of a third claw. (See Pl. 26ᵇ, Fig. 3.) There is often also a third line of tracks between the other two, as if produced by the tail or stomach of the animal touching the ground. Where the animal passed over the ridges of the ripple markings or wrinkles on the sand, they are flattened and brushed down. Thus a ridge between b. and d. (Pl. 26ᵇ, Fig. 3) has been flattened, and there is a hollow at c. on the steep side of the ridge, which may have been produced by the animal slipping down or climbing up the acclivity.

Fig. 8. *a.* Tube on the under surface of a scale for the passage of the mucous duct. See V. I. Note, p. 191, 192. (Agassiz.)

Fig. 9. Anterior extremity of the lower jaw of Holoptychus Hibberti, from the Lime stone of Burdie house, near Edinburgh. See Note, V. I. p. 275. The rugged surface of this bone is very remarkable. (Hibbert.)

Fig. 9′. Small teeth of Holoptychus Hibberti, fluted externally towards their base, and having a hollow cone within. (Hibbert.)

Fig. 9″. A small tooth magnified. (Hibbert.)

Fig. 10. One of the larger teeth in the Jaw of Holoptychus Hibberti, deeply fluted at the base, and having a hollow cone within. None of these teeth have sockets, but they adhere by a bony attachment to the jaw. (Hibbert.)

Fig. 11. Tooth of Holoptychus Hibberti. (Hibbert.)

Fig. 12. Tooth of Megalichthys Hibberti.* (Hibbert.)

Figs. 13, 14. Teeth of Holoptychus Hibberti, (Hibbert.)

Figs. 11. 12. 13. 14. are from Burdie house.

* Since the discovery of Megalichthys, which we have quoted in V. I. p. 276, Mr. W. Anstice, of Madeley, has found two jaws and punctate scales of the same species, in nodules of Iron stone from the Coal field of Coalbrook Dale; he has also found Ichthyodorulites, bones of fishes, and Coprolites, forming the nuclei of other balls of the same Iron stone.

Mr. Murchison has still more recently (1835) discovered remains of the Megalichthys, Holoptychus, and Coprolites, with several species of Unio, in the Wolverhampton Coal field. These great Sauroid fishes, which were first recognized at Edinburgh, in Sept. 1834, have also been detected in the English Coal fields of Newcastle on Tyne, Leeds, and Newcastle under Lyne.

PLATE 27ᵃ.   V. I. p. 274.

Fig. 1. Lepidosteus osseus, or bony Pike of North America. (Agassiz. Vol. 2. Tab. A.)

Fig. 2. Portion of the lower Jaw of Lepidosteus osseus, shewing the occurrence of a row of larger conical hollow teeth, fluted externally, between two rows of smaller Teeth. (Original.)

2. a. Longitudinal section of a large Tooth, shewing the internal hollow cone. (Original.)

2. b. Transverse section of a large Tooth. (Original.)

Fig. 3. Transverse section of the Jaw. fig. 2. (Original.)

Fig. 4. Fragment of a small upper Jaw of Megalichthys Hibberti, from Burdie house, shewing a disposition of large and small teeth, similar to that in fig. 2. (Hibbert.)

4. a. b. Transverse section of the larger teeth.

4. c. Longitudinal section of a large Tooth.*

4. d. Punctate scale of Megalichthys.

Fig. 5. Aspidorhynchus: a fossil Sauroid fish from the Lime stone of Solenhofen. (Agassiz, Vol. I. Tab. F.)

PLATE 27ᵇ.   V. I. p. 278.

Amblypterus: one of the fossil fishes peculiar to the Carboniferous strata. (Agassiz, Vol. I. Tab. A. fig. 3.)

---

* It appears that in the Megalichthys and Holoptychus the structure of the teeth, both large and small, was precisely the same as in the large and small teeth of Lepidosteus osseus, both as to the hollow internal conical cavity, and the external flutings towards the base, and also as to their mode of growth by ascent of fibrous matter from the bony substance of the jaw, and not from roots placed in deep alveoli, as in many of the Saurians.

PLATE 27ᶜ.   Vol. I. p. 281.

Fig. 1. Fossil fish of the genus Microdon, in the family
Pycnodonts. (Agassiz, Vol. I. Tab. G. fig. 3.)

Fig. 2. Os Vomer of Gyrodus umbilicatus, from the Great
Oolite of Durrheim, in Baden. (Agassiz.)

Fig. 3. Os Vomer of Pycnodus trigonus, from Stonesfield,
Oxon. (Original.)

PLATE 27ᵈ.   V. I. p. 287, Note.

A. Teeth of a recent Shark, allied to fossil species.

Fig. 1. Anterior and Palatal Teeth of the Port Jackson
Shark, (Cestracion Phillippi.) (Phillip.)

Fig. 2. Anterior cutting teeth of Port Jackson Shark, in
the College of Surgeons, London. (Owen.)

Fig. 3. Flat tessellated tooth of the same. Nat. size.
a. Outer articular facet, shewing the tubular structure of the bony base. *b.* Punctate surface of the
superficial enamel. (Owen.)

Fig. 4. Mesial, and inner articular facet of another large
tooth of the same. *a.* Upper concave margin thinly
covered with enamel. *b.* Lower bony margin without
enamel. *a'. b'.* Bony base of the tooth exposed by
removal of the Enamel. The surface is areolar,
from the bending and blending together of the bony
tubes. *c. c'.* Fractured edge of the marginal and
superficial enamel. (Owen.)

Fig. 5. Another anterior cutting tooth. *a.* Smooth ena-
melled point. *b.* Minutely rugous and tuberculated
base. In some of the cutting teeth both sides of the
base are rugous. (Owen.)

B. Various forms of fossil Teeth, in the three sub-fami-
lies of Sharks. (B. 1. to B. 13. Agassiz.)

Figs. 1—5. Teeth of fossil Sharks in the sub-family of
Cestracionts. See V. I. p. 287.

Fig. 1. Psammodus, from Mountain limestone, Bristol.

Fig. 2. Orodus, from the same.

Fig. 3. Acrodus, from the Lias, Lyme Regis.

Fig. 4. Ptychodus, (upper surface) from the Chalk.

Fig. 5. Side View of fig. 4.

Figs. 6—10. Teeth of extinct fossil Sharks in the sub-family of Hybodonts; in this family the enamel is *plicated* on *both* sides of the teeth. See V. I. p. 288, Note.

Fig. 6. Side view of tooth of Onchus, from the Lias at Lyme Regis.

Fig. 7. Front view of the same.

Figs. 8. 9. 10. Teeth of Hybodonts, from the Oolitic slate of Stonesfield, Oxon.

Figs. 11. 12. 13. Fossil Teeth of true Sharks in the Squaloid division of that family, having the Enamel *smooth* on the *outer* side. From the Chalk and London clay. See V. I. p. 289, Note.

Fig. 14. Palatal teeth of Myliobatis striatus, from the London clay of Barton cliff, Hants. See V. I. p. 291. Much of the enamel is worn away by use, as frequently happens in the tongue and palatal bones of living Rays. (Original.)

C. Petrified remains of an extinct Genus of Shark.

Fig. 1. Jaw of Hybodus reticulatus, from the Lias at Lyme Regis. (scale one half.) Many of the Teeth retain their place on the margin of the bone. The granulated structure of bone is distinctly preserved. (De la Beche.)

Fig. 2. Teeth selected from the Jaw last figured. Nat. size.

Fig. 3. Icthyodorulite, from the Lias at Lyme Regis, being the Dorsal spine of Hybodus incurvus, set with teeth-like hooks, to suspend the membrane of the dorsal fin. (De la Beche.)

A double row of similar hooks occurs on the first
dorsal ray of the Barbel, (Barbus Vulgaris.) And
on the anterior ray both of the dorsal and anal fins
of the Carp, (Cyprinus Carpio.)

Fig. 4. Transverse Section of fig. 3, at *a.** (De la Beche.)

PLATE 27°.   V. I. p. 288.

Fig. 1. Portion of the palatal teeth of Acrodus nobilis,
resembling a cluster of contracted Leeches. These
teeth are in their natural place, adhering to the
curved granular bone of the palate, which is well
preserved, and impregnated with Carbonate of lime.
(Miss S. C. Burgon. Original.)

Fig. 2. Continuation of the three rows of teeth on the
reverse of fig. 1.   Scale one half. (Original.)

Fig. 3. One of the largest teeth on the centre row,
having the upper part of the Enamel worn away
by friction. Nat. size. (Original.)

Fig. 4. Magnified view of the minute tubercles of Enamel
which grew upon the skin; the decay of the skin

---

* In the Lond. and Edin. Phil. Mag. Jan. 1836, the author has
published a notice of his recent discovery of the jaws of four extinct
species of fossil fishes of the genus Chimæra, a genus hitherto unknown in a fossil state. The only known species (C. monstrosa)
approximates most nearly to the family of Sharks; and is found
pursuing Herrings and other migratory fishes. The Chimæra is one
of the most remarkable among living fishes, as a link in the family of
Chondropterygians; and the discovery of a similar link, in the geological epochs of the Oolitic and Cretaceous formations, shews that
the duration of this curious genus has extended through a greater
range of geological epochs, than that of any other genus of fishes yet
ascertained by Professor Agassiz, and leads to important considera-
tions in Physiology.

The Chimæra partakes of one remarkable character with the Cestracion Phillippi, whereby this species alone, among living Sharks,
is connected with the extinct forms of that family, in having the first
ray of the dorsal fin enlarged into a strong bony spine armed with
sharp hooks, like the *Ichthyodorulite* of the earliest fossil Sharks.

has brought clusters of these tubercles into contact
with the bone in several parts of fig. 1. (Original.)

Fig. 5. Magnified view of similar minute tooth-like tubercles of Enamel, forming the Shagreen on the
skin of the head of the recent Squatina angelus.
See V. I. p. 269, Note. (Original.)

PLATE 27ᶠ. V. I. pp. 286 & 289.

Beautiful cluster of palatal teeth of Ptychodus polygyrus, from the Chalk. Insulated teeth of many species
of this Genus abound throughout the Chalk formation.
The mouth of these and all the other numerous extinct
species of Sharks in the family of Cestracionts, was lined
with a pavement of similar powerful teeth, forming a most
efficient apparatus, for crushing the shells of Crustacea
and Conchifera, which probably formed their principal food.
The surfaces of the Enamel are often worn away, like that
at Pl. 27ᵉ. fig. 3. The strength and efficacy of these teeth,
viewed as Instruments for crushing shells, is very remarkable. Beneath the Enamel, the body of each tooth is
composed of a strong mass of bone. (Miss F. C. Burgon.
Original.)

PLATE 28. V. I. p. 303.

Fig. 1. represents the common calmar or squid (Loligo
vulgaris, Lam. Sepia loligo, Linn.) shewing the
place and excretory duct of its Ink bag, and the
position of the feet on the anterior margin of the
head. (Blainville.)

Fig. 2. Side view of the Pen of the Loligo vulgaris,
shewing its position in the back of the animal,
fig. 1. (Original.)

Fig. 3. Concave under surface of the same pen. (Original.)

Fig. 4. Convex upper surface of portion of another recent pen, of the same kind. The structure of figs. 3 and 4 closely resembles that of the fossil species represented at fig. 6, of this same Plate, and also at Pl. 29. fig. 1. and Pl. 30. In all of them, the horny plates are composed of a series of longitudinal fibres, intersected by another series of transverse fibres. The disposition of the transverse fibres is most simple in the recent species; passing obliquely outwards from each side of the central shaft, like the barbs or fibrils in the vane of a feather, and being most distinct towards the outer margin.

The longitudinal fibres are scarcely visible in the recent species, except where they are collected into fluted fasciculi, (Pl. 28. fig. 4. BB.) in those parts which correspond with the *marginal bands* of the fossil species. (Original.)

C. Central part of the Pen, raised like the shaft of a quill between its fibrils.

Fig. 5. Ink bag of a recent Cuttle fish, dissected by the author at Lyme Regis, 1829, containing its natural Ink in a desiccated state; it is a black shining Jet-like substance, having a splintery fracture, and resembling the substance and fracture of the fossil Ink. Its bulk is not much reduced by desiccation. (Original.)

Fig. 6. Upper convex surface of a fossil pen of Loligo Aalensis from the Lias of Lyme Regis. A. A. the barbs; B. B. the marginal bands; C. axis of the shaft; D. Excretory duct of the Ink bag, distended with petrified Ink.* (Original.)

---

* In this specimen we see distinctly the disposition of the marginal bands.

Fig. 7. Upper surface of Fossil Loligo from the Lias of
　　　Lyme Regis. A, A, Barbs of the Pen. B, B, Mar-
　　　ginal bands. C, Axis of the Pen. d, upper plate
　　　of marginal band, having an unusually corrugated
　　　surface, which may be the result of imperfect growth
　　　of the transverse fibres; if fully expanded they would
　　　probably have resembled those of the subjacent
　　　Plate at d'. (Original.)

d'. Magnified representation of the rugous surface of d.

d'''. Magnified representation of the second plate of the
　　　marginal band, Fig. 7. d'.

e. Upper surface of second Plate of the shaft of the pen;
　　　here the transverse wavy lines predominate over
　　　the vertical straight lines; but both are visible.

f. Upper surface of third plate; here the vertical straight
　　　fibres prevail over the transverse wavy fibres.

PLATE 29.　V. I. pp. 307 and 309.

Fig. 1. Fossil Loligo from Lias at Lyme, in the collec-
　　　tion of Miss Philpot, exhibiting nearly the same
　　　structure as figs. 6. 7. at Pl. 28. and containing be-
　　　neath the pen, a very large Ink bag, D. The greater
　　　proportionate size of this Ink bag indicates a dif-
　　　ference in species from fig. 3. (Mrs. Buckland.
　　　Original.)

Fig. 2. Loligo Aalensis from Lyme Regis shewing the
　　　under surface or concave side, and the duct of the
　　　Ink bag distended with Ink. A.A. Barbs or fila-
　　　ments of the Pen; B.B. Marginal bands; C. Axis
　　　of Shaft; D. Duct of Ink bag. (Mrs. Buckland.
　　　Original.)

　　　The wavy lines here seen between the Ink bag
　　　and the apex of the Pen, are the inferior termina-
　　　tions of the successive laminæ of growth; each

larger and superior Plate overlapping the edges of
the next subjacent and smaller plate. These edges
are rendered more irregular by decomposition.

d. Magnified representation of very minute curved lines
passing from the marginal band across the shaft,
at d.

e. Thin lamina of the white pulverulent substance of a
decomposed Plate; it retains partial traces of the
transverse wavy fibres.

f. Minute perpendicular filaments prevailing over the
transverse fibres of the shaft.

Fig. 3. Fossil Loligo from Lyme Regis, shewing the same
structure as the preceding figures, in the several
portions of the Pen that are preserved; and having
its Ink bag distended nearly in its natural shape and
place beneath the Pen. (Original.)

C. C. Axis of the shaft.

Figs. 4. 5. 6. 7. 8. 9. Fossil Ink bags from Lyme Regis.
The membranous sacs and excretory ducts are still
preserved, and closely resemble those of a recent Ink
bag; see Pl. 28. fig. 5. (Original.)

Fig. 10. Fossil ink bag found by Miss Anning in the
Lias near Watchet, Somerset. (Original.)

PLATE 30.   V. I. p. 309.

A large fossil pen of Loligo; from the Lias at Lyme Regis.
In the collection of Miss Philpot. (Mrs. Buckland.
Original.)

AA. Barbs of the pen, proceeding from the outer edges
of the marginal bands.

B. B. Marginal bands dividing the bases of the barbs
from the internal part or body of the shaft.

C. Axis of the Pen, dividing the body of the shaft into
two equal parts.

D. Transverse section across the Ink bag.

d. First or upper plate. This plate is very thin, and
smooth, and its structure is obscure, except on the
right marginal band at d', where the longitudinal
ridges on its surface are very distinct.

e. Upper surface of second plate, marked with broad
wavy lines, passing on each side from the axis outwards, across the body of the shaft, and over the
marginal bands.

f. Upper surface of a third plate, exhibiting minute
curved striæ, ascending symmetrically in opposite
directions from each side of the axis of the shaft C,
and descending towards its margin. These curved
striæ are intersected by minute longitudinal straight
lines, running nearly parallel to the axis of the
shaft. Towards the apex of the shaft at f', the broad
transverse curves predominate over the fine longitudinal fibres which lie beneath them. At g, no
transverse curves are visible.*   (Mrs, Buckland.
Original.)

PLATE 31. V. I. p. 317.

Fig. 1. Animal of Nautilus Pompilius, fixed in its shell.
The shell is copied from one in the collection of Mr.
W. I. Broderip. (Animal from Owen. Shell ori-
ginal.)

n. The Hood, or ligamento-muscular disk that surrounds
the head.

p. The digital tentacles protruded from their sheaths.

k. Funnel.

a. b. c. d. e. Siphuncle. The desiccated membrane of

* Herman von Meyer (Palæologica, 1832, P. 322,) mentions the
occurrence of ink bags, together with the horny internal shells of
Sepia, (Onychoteuthis) in the Lias of Culmbach and Banz.

the siphuncle is laid bare at *a. b. c. d.* At *e, e,* and
from thence inwards, it is covered by a soft calcareous coating or sheath.

y. y. Collar, projecting inwards from the transverse
plates, and supporting the Siphuncle. See Note,
V. I. p. 322.

Fig. 2. Upper horny mandible of the animal, with a hard
calcareous point. (Owen.)

Fig. 3. Lower horny mandible, armed with a similar
calcareous point. (Owen.)

Fig. 4. Calcareous point, and palate of upper mandible
separated from the horny portion. (Owen.)

Fig. 5. Under surface, or palate of a Rhyncholite, or
fossil beak, from the Lias at Lyme Regis, analogous
to the recent specimen, fig. 4. (Original.)

Fig. 6. Upper view of another Rhyncholite from the same
stratum and place. Black portions of the horny
substance, in a state resembling charcoal, remain
attached to its posterior surfaces. (Original.)

Fig. 7. Side view of the calcareous portion of an upper
mandible, from the Muschel kalk of Luneville.
(Original.)

Fig. 8. Upper view of another Rhyncholite from Luneville. (Original.)

Fig. 9. Palatal view of fig. 8. (Original.)

Fig. 10. Calcareous point of an under mandible from
Luneville. The dentations on its margin resemble
those on the recent mandible, fig. 3, and co-opera-
ting with the dentations on the Margin of the upper
mandible, fig. 9, must have formed an Instrument
(like the recent beak, figs. 2 and 3,) well fitted for
the rapid demolition of Crustacea and small Shells.
(Original.)

Fig. 11. Under surface of fig. 10.; it is strengthened by

a double keel-shaped indented process, enlarging
from its apex backwards.* (Original.)

PLATE 32. V. I. p. 323.

Fig. 1. Part of the petrified shell, and casts of the
interior of some of the chambers, of a Nautilus
hexagonus, from Marcham, Berks. This fossil exhibits at its smaller End, from *d.* to *b.*, a series of
casts of the Air chambers, from which the external
shell has been removed. The cavity of each chamber is filled with a disc of pure calcareous spar,
representing the exact form of the chamber into
which it had been infiltrated. In the larger portion
of this fossil, the petrified shell retains its natural
place, and exhibits fine wavy lines of growth forming minute Ribs across its surface. (Original.)

Fig. 2. Fractured shell of N. hexagonus, from the Calcareous grit of Marcham. The chambers are lined
with calcareous spar, and a circular plate of
the same spar is crystallized around the siphon.
The interior of the siphon is filled with a cast of
Calcareous grit, similar to that which forms the
rock from which the shell was taken. See V. I.
p. 326.† (Original.)

* Although the resemblances between these fossil beaks, and that
of the animal inhabiting the N. Pompilius, are such as to leave no
doubt that Rhyncholites are derived from some kind or other of
Cephalopod, yet, as they are found insulated in strata of Muschel
kalk and Lias, wherein there occur also the remains of Sepiæ that
had no external shells, we have not yet sufficient evidence to enable
us to distinguish between the Rhyncholites derived from naked Sepiæ,
and those from Cephalopods that were connected with chambered
shells. I possess a specimen of a fossil Nautilus from the Lias at
Lyme Regis, in which the external open chamber contains a Rhyncholite.

† This fossil exhibits the Siphuncle in its proper place, passing

Fig. 3. represents in its natural size, a portion of the
Siphuncle which in Fig. 2. is laid bare along its
course through the chambers, *d. e. f.* In the transverse Plate, *h*, the siphuncular collar is entire, but
a Section of another collar in the transverse Plate,
i, shews the contraction of the Siphon at its passage
through this aperture, and exhibits also the overlapping, or squamous suture by which the Collar
is fitted to the superior and inferior portions of the
calcareous Sheath of the Siphon. See V. I. pp.
326, 327. Note. (Original.)

A similar structure may be seen at the Collars
of the transverse Plates of the N. Striatus. See
Pl. 33.

across the cavities of the Air chambers. As in the recent Nautilus
Pompilius, there is no communication between the interior of the
Siphon and that of the Air chambers, so in this fossil shell, there is
proof that no communication existed between these cavities. A
transverse section at *a.* shews the thin edge of the sheath of the siphuncle, surrounded externally with calcareous spar, and filled
internally with Grit. Other Sections of the Siphuncle at *b. d. e. f.*
shew the calcareous Grit within its cavities to be contracted at its
passage through the collars of the transverse plates, and most enlarged midway between one transverse plate and another.

This fossil affords two proofs that no communication existed between
the interior of the Siphuncle and that of the Air chambers. 1st.
the calcareous sheath of the Siphuncle is seen at *d. e. f.* completely
enclosing the calcareous grit which forms the cast within it. 2dly,
had there been any communication between the interior of the siphuncle, and that of the air chambers, these chambers must have
received some portion of the materials of the grit that have filled
this Siphuncle: not a particle of grit is found in any one of the
adjacent air chambers, but they are all lined, and some of them
nearly filled with a crystalline deposit of Carbonate of Lime, disposed
in uniform plates around the interior of each chamber, and around
the Siphuncle. *See Fig. 2. c. c¹. a. a¹. a². a³. and Fig. 3. d—h.* This
deposit can only have been formed from water charged with carbonate of lime, introduced by infiltration, after the interment of the
shell, and filling the chambers which are thus uniformly invested.

PLATE 33.   V. I. pp. 326, 327. Note.

Longitudinal Section of Nautilus Striatus, from the Lias
at Whitby, in the collection of Mrs. Murchison.   The interior of the Chambers is filled exclusively with calcareous
spar, and that of the Siphuncle with Lias.   (Original.)

    a. The Siphuncle: the union of the siphuncular calca-
        reous sheaths, with the aperture or collar of each
        transverse Plate, is so closely fitted, that no fluid
        could have passed between them into the air cham-
        bers.

    b. One of the transverse Plates forming the Air cham-
        bers.

    c. White calcareous spar, filling the middle region *only*
        of the air chambers.

    d. Stratified zones of dark coloured calcareous spar, de-
        posited in equal thickness on both sides of the
        transverse plates, and also on the inside of the shell,
        and around the calcareous sheath of the siphuncle.*

    e. Portion of the external shell, shewing a laminated
        structure.

PLATE 34.   V. I. p. 329. Note.

Drawing of the animal of the Nautilus Pompilius, pre-
pared at my request by Mr. Owen, to shew the manner
in which the siphuncle terminates in the Pericardium.
(Original.)

   * The successive zones of this dark Spar shew that the Lime composing it was introduced by slow and gradual infiltrations into the
cavity of the air chambers.   Hence it follows that no communication
existed between the Siphuncle and these chambers, at the time when
this Pipe was filled with the fluid mud, that has formed a cast of Lias
within it.   As the fractures across the Siphuncle in the 2nd and 3rd
chambers are filled only with spar, of the same kind as that within
these Chambers, these fractures could not have existed, when the
Mud of the Lias formation entered the Siphuncle, without admitting
it also into the chambers adjacent to them.

a. The Heart.

b. A bristle passing from the pericardium through the membranous siphuncle laid bare.

c. Bristles passing from the pericardium through the orifices of communication with the Branchial chamber.

d. d. d. d. Follicles communicating with the Branchial Arteries.*

'd. 'd. 'd. 'd. Pericardial septa, forming thin muscular Receptacles of the follicles.

e. e. The Branchiæ.

f. The Branchial Chamber.

g. The Funnel, or Branchial outlet.

h. The infundibular valve.

i. i. The digital processes.

k. The Gizzard.

l. The Ovary.

m. m. The mantle dissected off.

n. The membranous siphuncle.

o. o. The siphuncular artery.

p. p. The Boundaries of the Pericardial cavity.

q. Portion of the Siphuncle between the Pericardium and first transverse plate of the shell.†

---

* Mr. Owen supposes that these follicles discharge the impurities of the blood into the Pericardium, when there is no access of water to the Branchiæ, during the time that the animal is contracted within its shell. The overflowings of this pericardial fluid may pass out through the orifices marked by the bristles, c. c.

† This upper portion or neck of the Siphuncle, has the form of a flattened canal, with thin Parietes of the same substance as the Pericardium ; when the animal expands itself at the bottom of the sea, this neck is probably closed by the lateral pressure of the gizzard, *k,* and ovary, *l,* and so acts instead of a valve to prevent the return of the pericardial fluid into the Siphuncle. At such times, the deep-sea water must press with great force on the exterior of the Pericardium, and tend to force the pericardial fluid into the Siphuncle; but as an equal amount of pressure is applied simultaneously to the Ovary and

PLATE 35.   V. I. p. 339.

Cast of the interior of the Shell of Ammonites obtusus from Lyme.   Fragments of the shell remain near *b*. and *e*.

One object of this Plate and of many of the figures at Pl. 37. is to shew the manner in which the external shell is fortified by Ribs and Flutings, (PP. 340. 341.) and further supported by the edges of the internal transverse plates, that form the air chambers. See V. I. p. 348, Note. (Original.)

PLATE 36.   V. I. p. 338. Note.

Longitudinal section of another shell of Ammonites obtusus from the Lias at Lyme Regis.   (Original.)

The greater part of the outer chamber, and the entire cavities of the air chambers are filled with calcareous spar, and the Siphuncle, (preserved in a carbonaceous state,) is seen passing along the entire dorsal margin to the commencement of the outer chamber. See V. I. p. 351, Note.

Von Buch has found evidence to shew that the membranous siphuncle of Ammonites was continued to a considerable distance along the outer chamber, beyond the last or largest transverse Plate. This discovery accords with the analogies afforded by the membranous neck of the siphon of the N. Pompilius, which is continued along the outer chamber from the last transverse Plate to the Pericardium. See Pl. 34. *q*.*

Gizzard, the lateral pressure of these two organs on the neck of the Siphuncle would tend to close it with a force exactly counterbalancing the external pressure on the Pericardium.

* As the body of the animals that inhabited the Ammonites was more elongated than that of those inhabiting of the shells of Nautili, in consequence of the smaller Diameter of their outer Chamber, the place of their Heart was probably more distant from the last transverse Plate, than that of the Heart of Nautili; and the membranous Siphon connected with the Pericardium consequently longer.

PLATE 37. V. I. p. 341. Note.

Figs.		Locality.	Stratum.
1. Ammonites Amaltheus Gibbosus . . . . . (Schlotheim) .		Gloucester . .	Lias.
2. A. Varicosus . . . . (Sowerby)		Black Down, Devon.	Green Sand.
3. A. Humphriesianus . (Sowerby) . .		Sherborne . .	Inferior Oolite.
4. A. Lamberti . . . . . (Sowerby) . .		Oxford	Oxford Clay.
5. A. Planulatus . . . . (Schlotheim) .		Franconia . .	Jura limestone.
6. A. Bucklandi . . . . (Sowerby) . .		Bath	Lias.
7. A. Lautus . . . . . . (Sowerby) . .		Folkstone. . .	Gault.
8. A. Catena . . . . . . (Sowerby) . .		Marcham. . .	Calcareous Grit.
9. A. Variana . . . . . (Zieten) . . .		Geislingen . .	Jura limestone.
10. A. Striatus . . . . . (Reinicke) . .		Gros Eislingen	Lias.

    a. Exterior dorsal margin.
    b. Back view of the shell.
    c. Transverse section of shell.

The figures in this Plate are selected to exemplify some of the various manners in which the shells of Ammonites are adorned and strengthened by ribs, and flutings, and bosses. In Vol. I. p. 340, instances are mentioned of similar contrivances which are applied in Art to strengthen thin plates of metal. Workers in Glass have also adopted a similar expedient in their method of fortifying small wine flasks of thin glass, made flat, and portable in the pocket, with a series of spiral flutings passing obliquely across the sides of the flask, as in many of the flattened forms of Ammonite. Similar spiral flutings are introduced for the same purpose on the surface of thin glass pocket smelling-bottles. In other glass flasks of the same kind which are made in Germany, the addition of bosses to the surfaces of the flat sides of the bottles, produces a similar double result of ornament and strength.

PLATE 38. V. I. p. 347. Note.

Air chambers of Ammonites heterophyllus, filled with Lias, and shewing in a remarkable degree the effect of the undulating course of the edges of the transverse plates beneath the flat sides of the outer shell.

A portion of the outer shell is preserved at *c.* and impressions of the fluted interior of the shell, which has fallen off, are visible at *d.* (Original.)

PLATE 39. V. I. p. 348. Note.

This Plate presents a longitudinal view of the same fossil, of which a side view is given in the last figure. The same transverse plates; that approximate so closely beneath the sides of the shell, where it is flat and feeble, (Pl. 38.) are distant from each other along the dorsal portion, which from its convex form is strong.

The Siphuncle is preserved in its proper dorsal place at *d.*

The elevations and depressions of the transverse plate in front of this figure exemplify the theory of Von Buch, respecting the use of the Lobes and Saddles formed by the undulations of its outer margin. See V. I. p. 353, and Note. (Original.)

PLATE 40. V. I. p. 360. Note,

Fig. 1. Ammonites Henslowi (Goniatites), from Transition lime-stone in the Isle of Man.

The Lobes are simple, and without foliations; their form resembles that of the slipper-shaped lobe of the Nautilus Ziczac, and Nautilus Sypho. See Pl. 43.

The lobes D. L. l. V. are *pointed* inwards, and the intermediate Saddles S. d. S. L. S. V. are *rounded* outwards; according to the type of Ammonites. (Original.)

Fig. 2. Ammonites striatus (Goniatites), from the Coal Shale of Lough Allen in Connaught, having its lobes and saddles disposed in the same directions as in Fig. 3, the delicate longitudinal striæ and transverse ribs of the outer shell are strengthened by repeated intersections of the subjacent edges of the transverse Plates. (Original.)

Fig. 3. Back view of Ammonites sphæricus, from the limestone of Derbyshire, shewing the position of the siphuncle upon the dorsal margin, with its collar advancing outwards between the two simple dorsal lobes ; the lateral lobes are also simple and without foliations, and pointed inwards. (Martin Pet. Der. T. 7.)

Fig. 4. Ammonites nodosus (Ceratites). This is one of the species peculiar to the Muschel Kalk. The descending lobes terminate in a few small denticulations, *pointed* inwards, and the ascending saddles are *rounded* outwards, after the normal character of Ammonites. (Zeiten. Tab. II. Fig. 1. a.)

Fig. 5. Back of A. Nodosus, shewing the dorsal lobes *pointed* inwards, and the collar around the siphuncle advancing outwards. No edges of the transverse plates are placed beneath the dome-shaped Tubercles; these derive sufficient strength from their vaulted form. (Zeiten. Tab. II. Fig. 1. b.) .

PLATE 41. V. I. p. 349.

Ammonites giganteus, found in the Portland stone at Tisbury in Wiltshire. This beautiful fossil is in the collection of Miss Benett. The chambers are all void, and the transverse Plates and Shell converted to Chalcedony. (Original.)

PLATE 42. V. I. pp. 350, 351. Note.

Fig. 1. Cast of a single chamber of Nautilus hexagonus, shewing the simple curvatures of the edges of the transverse plates, and the place of the Siphuncle. (Original.)

Fig. 2. Cast of a chamber of Ammonites excavatus, having a complex form derived from the denticulated edges of the transverse plates. See V. I. pp. 350, 351, Note. (Original.)

Fig. 3. Casts of three chambers of Ammonites catena, with the Membrane of the Siphuncle on its dorsal margin. See V. I. p. 350, Note, and p. 351, Note.

The course of the transverse plates is beneath the *depressed* and weakest parts of the external shell, avoiding the bosses at c, d, e, which from their form are strong. (Original.)

Fig. 4. Ammonites varicosus, from the Green Sand of Earl Stoke, Wilts. Nat. size. See V. I. p. 351, Note. (Original.)

Figs. 5. 6. Portions of the same shell, having the transverse Plates and Siphuncle converted to Chalcedony. See V. I. pp. 351 and 352, Note. (Original.)

Fig. 7. Ammonites variocostatus, (nobis,) an undescribed species of Ammonite from the Oxford Clay at Hawnes, 4 m. S. of Bedford.   Diameter 9 inches.

The name Variocostatus expresses the remarkable change in the character of the Ribs, near the outer termination of the air chambers.

On the inner whorls of the shell, these ribs are narrow, and highly raised, set close to one another, and bifurcated at the back of the shell, (from *d.* to *c.*); but near the outer chamber (*b.* to *a.*) they become broad and distant, and the dorsal bifurcation ceases.

The edges of the transverse plates are exposed by the removal of the shell from *c.* to *b.*, they appear also at *a. d.* (Original.)

Similar variations in the form of the ribs occur in Ammonites biplicatus and Ammonites decipiens.

PLATE 43. V. I. pp. 358, 359.

Fig. 1. Fragment of Nautilus sypho, in the collection of W. I. Broderip, Esq. from the Miocene division of the Tertiary formations at Dax, near Bourdeaux. The accidental fractures of this fossil afford an instructive display of the disposition of the transverse Plates and Siphuncle. (Original.)

Fig 2. Another fractured shell of the same species from Dax, in the collection of Mrs. Buckland, shewing at a^1, a^2, a^3, the disposition of the lateral lobes. See V. I. p. 359, Note. (Original.)

Fig. 3. Cast of the interior of Nautilus Ziczac, in the collection of Mr. James Sowerby, shewing the disposition of the lateral lobes. (See V. I. pp. 359, 360. (Original.)

Fig. 4. Cast of a single chamber of Nautilus Ziczac, in the collection of Mr. J. Sowerby, shewing the disposition of the ventral and dorsal Lobes and Siphuncle. See V. I. p. 359, Note. (Original.)

PLATE 44. V. I. p. 361, et seq.

Fig. 1. Molluscous animal inclosing the Spirula Peronii. See V. I. p. 362.* (Blainville.)

Fig. 2. Section of a Spirula (Nat. size), shewing its transverse Plates and siphuncular sheath. (Original.)

* M. Robert has recently discovered between the Canaries and Cape Blanc, several imperfect bodies of a small species of molluscous animal, each inclosing a Spirula.

In all these the position of the shell is not at the posterior extremity, as in the figure of the specimen found by Peron, but in the back, parallel to the axis of the body, like the shell of the Sepio-taire, or internal shell of the common Sepia. This position agrees with that of the animal figured by Blainville, if we suppose the caudal portion of the latter to have been lost.

On each side of the body are two expansions that act like Fins, as in the Sepiole. Beneath the neck is the aperture of the Funnel.

Fig. 3. Lituite in the Transition lime-stone of Oeland.

a. Siphuncle of Lituite. (Original.)

Fig. 4. Section of an Orthoceratite in the Transition lime-stone of Oeland, in the Collection of C. Stokes, Esq. (Original.)

a. Siphuncle of the same.

Fig. 5. Baculite, from Chalk of the Cotentin; terminating at its large end in the chamber *a.* (Original.)

Fig. 5. *b.* Front view of the transverse plate of a Baculite, shewing the margin to be disposed in lobes and saddles, and the place of the Siphuncle to be on the back of the shell at *c.* (Original.)

Fig. 6. Transverse section of a Nummulite. (Parkinson, V. 3. Pl. X. Fig. 16.)

Fig. 7. Longitudinal section of another Nummulite.* (Parkinson.)

In one specimen the Eye is preserved, and is very large in proportion to the body. These Mollusks form the prey of the Physali, and were caught entangled in their Tentacula.

L'Echo du Monde Savant, 1 Mai, 1836.

* Among the microscopic fossil shells placed by D'Orbigny in the same Order as Nummulites (*Foraminiféres*), Count Munster enumerates 40 species from the Cretaceous free stone of Maestricht. Mr. Lonsdale also has discovered 16 species of microscopic foraminifers in the English Chalk. (See V. I. p. 448, Note.) Microscopic shells of this Order occur in countless myriads throughout the Tertiary strata. (See V. I. p. 385.)

The Sand of the Shores of the Adriatic, and of many Islands in the Archipelago, is crowded with recent microscopic shells of the same kind.

It is mentioned in our Note, V. I. p. 382, that doubts have arisen as to the supposed origin of many of these minute multilocular shells from Cephalopods. Some recent observations of M. Dujardin have induced him to refer the Animals which construct the Miliola and some other microscopic foraminiferous shells, to a new Class of animals of lower degree than the Radiata, and possessing a locomotive power by means of minute tentacular filaments. He proposes to give them the name of Rhizopodes. Ann. des Sci. Nat. Mai, 1835. p. 312.

Fig. 8. Hamites Bucklandi, (Phillips,) from the Gault or Speeton Clay, in the collection of Mr. I. Phillips, of York. (Original.)

Fig. 8ª. Transverse septum of Fig. 8, shewing the lobes and saddles, and the siphuncle at *a*.

Fig. 9. Hamites armatus, from the upper Green Sand, near Benson. (Sowerby.)

Fig. 10. Transverse section of the same, shewing the siphuncle, on the back, between the spines.

Fig. 11. Hamites from Folkstone Clay, shewing the spiral Ribs of the outer shell. At *a*. we see the Siphuncle, and the lobes and saddles of the transverse Plate.

Fig. 12. Fragment of the cast of the interior of another Hamite from Folkstone Clay, shewing the Siphuncle at *a*. The removal of the outer shell shews the sinuous edges of the transverse Plates beneath the Ribs. (Original.)

Fig. 13. Fragment of Hamites articulatus (Sow.) from the Green Sand at Earl Stoke, shewing the Siphuncle (*a*.) covered by a small portion of the shell. The sinuous terminations of the transverse plates are visible beneath the ribs, having their secondary lobes *rounded* outwards (*b*.) and *pointed* inwards (*c*.) like the secondary lobes of Ammonites. (Original.)

Fig. 14. Fragment of Turrilites Bergeri, in the collection of G. B. Greenough, Esq. from the Green Sand formation. The siphuncle is seen near the upper or dorsal margin of two whorls at *a. a.*; the sinuous edges of the transverse plates are visible on the middle whorl; and the entire surface of a transverse plate is laid open at the smaller end of a third whorl, shewing its lobes and saddles to be analogous to the same parts in Ammonites. (Original.)

Fig. 15. Scaphites Equalis, from Chalk near Rouen, in the collection of Mr. J. Sowerby; the sides of the external shell are strengthened and ornamented by ribs and tubercles; and the edges of the transverse plates disposed in sinuous foliations (c.) as in Ammonites. The mouth or outer margin (b.) returns so nearly into contact with the air chambers (c.), that the want of space at this part for the expansion of arms and head, makes it probable that the Scaphite was placed entirely within the body of its animal. (Original.)

Fig. 16. Transverse section of the chambered portion of Fig. 15, shewing the arrangement of the lobes and saddles to be similar to that of Ammonites; the siphuncle also is seen on the dorsal margin at a. (Original.)

Fig. 17. Longitudinal section of the calcareous Sheath and Alveolus of a Belemnite.

a. Alveolus, or internal shell, divided by transverse Septa into air chambers. See V. I. p. 373.

b. Siphuncle, passing along the margin of the air chambers.

c. Apex of the fibro-calcareous sheath, or solid Cone of the Belemnite.

PLATE 44'. V. I. p. 371, et seq.

Illustrations of the probable nature of the Animals that gave origin to Belemnites.*

---

* In the descriptions of Pl. 44'. and Pl. 44". the following letters indicate the same parts in each specimen to which they are applied.

a. The Apex of the calcareous shell, or sheath.

b. Alveolar portion, or chambered shell.

c. Ink-bag.

d. ⎫ Portions of the thin anterior horny sheath, sometimes highly
e. ⎬ nacreous.

f. Neck of Ink-bag.

Fig. 1. Imaginary restoration of Belemnosepia, shewing
the probable place of its Ink-bag, and of the internal
shell or Belemnite. The three component parts of
this Belemnite are represented as if longitudinally
bisected: the place assigned to this Ink-bag is
nearly the same as in the recent Loligo. (Original.)

Fig. 2. Sepia officinalis, shewing the position of the
internal shell or sheath (Sepiostaire) within the
dorsal portion of its sac. Its apex (*a,*) and calcareous dorsal plates (*e,*) correspond with the apex
calcareous conical sheath of a Belemnite.

Fig. 3. Sepia officinalis, laid open along the ventral
portion of its Sac, to shew the position of its Ink-
bag. (Original.)

Figs. 3. *a.* 3. *b.* 3. *c.* Rhyncholites, found in contact with
Belemnites in the Lias at Lyme Regis. Nat. size.
(Original.)

Fig. 3. *d.* Beak of a small Testudo from Chalk, in the
collection of Mr. Mantell, shewing a fibro-cancel-
lated bony structure, very different from the compact shelly condition of the Rhyncholite, for which
it may from its size and shape be mistaken. (Original.)

Fig. 4. Ventral surface of a Sepiostaire; the elongated
shallow cone, or cup, (*e. e. e'. e'.*) is composed of
very thin calcareous plates, alternating with horny
membranes, which are expanded outwards to form
the thin margin of the cone. This irregular cone
or shell represents the hollow cone at the larger
extremity of the Belemnite, (Fig. 7. *b. b'. e. e'. e''.*)
which includes its Alveolus (*b. b'.*) and Ink-bag (*c.*).
Within this shallow sub-conical shell of the Sepiostaire is contained its alveolus, or calcareous chambered portion, (Fig. 4. *b.*) which represents the

chambered alveolus in the Belemnite, (Fig. 7. *b. b'.*)
but has no Siphon.   (Blainville.)

Fig. 4'. Longitudinal section of the apex of the shell of
Sepia officinalis. This apex is composed of granular
calcareous matter (*a.*), alternating with conical horny
laminæ, which expand laterally into the horny margin (*e.*).   (Original.)

Fig. 5. Longitudinal view of Fig. 4.   The apex (*a.*) represents the apex of a Belemnite.   The back of the
shell (*e.*) the dorsal part of a Belemnite ; and the
alveolar portion (*b. b'.*) represents the internal chambered shell of a Belemnite.   (Blainville.)

Fig. 6. Anterior extremity of the lamellæ, or alveolar
plates, exposed by a longitudinal section in Fig. 5.
In the mature animal these lamellæ are nearly 100
in number ; a few of them only are here represented.

These alveolar plates form the internal chambers
of the Sepiostaire, and represent the transverse
plates of the Alveolus in Belemnites, and other
chambered shells; but as the Sepiostaire has no
siphuncle, its chambers seem not subservient, like
those of the Belemnite, to the purpose of *varying*
the specific gravity of the animal; the intervals between its plates are occupied by an infinite number
of thin winding partitions standing perpendicularly
between the lamellæ.

Figs. 6'. 6''. Thin calcareous partitions winding between,
and supporting the alveolar plates of the Sepiostaire.
The sinuous disposition of these partitions increases
their efficacy in resisting pressure, on the same
principle, as in the foliated edges of the transverse
plates of Ammonites.*   The sinuosity of the cal-

* Dr. Fleming has accurately described the structure of these
partitions, as exhibiting perpendicular laminæ, waved and folded in
brain-like gyrations which occasionally anastomose.

careous partitions is least near the margin of the lamellæ. See Fig. 6'. (Original.)

Fig. 6'''. Columnar appearance of the sinuous partitions when viewed laterally. (Original.)

Fig. 7. Unique specimen of Belemnites ovalis, from the Lias at Lyme Regis, in the collection of Miss Philpotts. A fracture at *b'*. shews the chambered areolæ of the Alveolus. At *e.* the thin conical anterior horny sheath originates in the edge of the calcareous sheath, and extends to *e''*. The surface of this anterior sheath exhibits wavy transverse lines of growth; it is much decomposed, slightly nacreous, and flattened by pressure.

Within this anterior conical sheath the Ink-bag is seen at *c.* somewhat decomposed, and partially altered to a dark grey colour. (Original.)

Fig. 8. Portion of the Ink-bag broken off from Fig. 7. *c.* and covered by that portion of the horny case which lay above it. The transverse lines, *e.* on this portion, are the continuation of the lines of growth on the horny sheath of Fig. 7. *e. e'. e''*. (Original.)

Fig. 9. Belemnites Pistilliformis? from the Lias at Lyme, in the collection of Miss Philpotts, having a portion of its ink-bag at *c.* (Original.)

Figs. 10. 11. 12. Belemnites from the Jura limestone of Solenhofen, figured by Count Munster in Boué's Mémoires Géologiques, Vol. I. Pl. 4. In 10 and 12 the form of the anterior horny sheath is preserved, to a length equal to that of the calcareous shaft of the Belemnite, but in none of them is the Ink-bag visible.* (Munster.)

---

* Von Meyer mentions (Palæologica, P. 322, 1st Edit. 1832,) that he has seen an Ink-bag at the upper end of a Belemnite from the Lias of Banz, and asks, "Do Belemnites possess an Ink-bag like that of the Sepia?"

Fig. 13. Chambered alveolar cone and horny sheath of a
large Belemnite from the limestone of Solenhofen;
the calcareous sheath or Belemnite itself has dis-
appeared. (Munster.)

Fig. 14. Belemnites brevis? from the Lias at Lyme;
Nat. size. The length of the shaft of this Belemnite does not exceed that of the Beloptera (Fig. 15);
a small fragment only of its alveolus is preserved,
but the place it occupied is filled with calcareous -
spar, and the hollow cone above it with lias. (Original.)

Fig. 15. Beloptera. In this fossil we have an intermediate
link between the Belemnite and the shell or sheath
of Sepia officinalis. *a.* represents the apex of the
sheath, *e. e.* its posterior expansion, analogous to
that at Fig. 4. *e. e.* and at Fig. 4′. *e.*; *e′* is its anterior expansion, bearing on its internal surface an-
nular marks derived from the transverse septa of the
alveolus. (Blainville.)

PLATE 44″. V. I. p. 374. Note.

All the figures in this Plate are of nat. size.

Fig. 1. Anterior Sheath and Ink-bag of Belemno-sepia,
discovered by Miss Anning in 1828 in the Lias of
Lyme Regis, and noticed by Dr. Buckland (Lond.
and Edin. Phil. Mag. May, 1829, P. 388,) as "derived from some unknown Cephalopod, nearly allied
in its internal structure to the inhabitant of the
Belemnite." This sheath is, for the most part, nacreous; in some places (*d. d.*) it retains the condition of horn. The corrugations on its surface indicate the lines of growth. At *f.* a transverse fracture
shews the neck of the ink-bag. (Original.)

Fig. 2. The lower part of Fig. 1. seen from another side;

the circular lines on the surface of its horny membrane *d*, are lines of growth. (Original.)

Fig. 3. Belemno-sepia from the Lias at Lyme, in the
Oxford Museum; the Ink-bag is preserved entire
within the anterior conical sheath *e. e. e.*; the
greater part of this sheath is highly nacreous, in a
few places (*d.*) it is horny. (Original.)

Fig. 4. Large Ink-bag from the Lias at Lyme, in the
collection of Mrs. Murchison, bearing on its surface
undulating lines of growth similar to those on the
surface of Fig. 1. The Ink is exposed at *c. c.*; in
other parts it is surrounded by the sheath, *e. e. e.*
Nearly one-half of this sheath retains the appearance of horn, whilst the other half is highly nacreous. This interchange of condition, from horn
to brilliant nacre, occurs in almost every specimen
from the Lias at Lyme, in which the Ink-bag is
accompanied only by the flexible anterior sheath,
and the calcareous sheath has perished. (Original.)

Figs. 5. 6. 7. 8. Ink bags from the Lias at Lyme, partially surrounded by brilliant nacre. In no one of
the specimens represented in Pl. 44″ is the least
trace of the calcareous sheath of the Belemnite preserved. See V. I. p. 376, Note. (Original.)

PLATE 45.* V. I. p. 392 et seq.

Fig. 1. Limulus Americanus (Leach), a young specimen from Honduras, one third of nat. size. *b′.*
Right compound Eye magnified. *b″.* Two single

---

* The following letters are applied in Pl. 45 and Pl. 46, to corresponding parts of different animals. *a.* the shield; *a′.* lateral
portion of the shield; *b.* the eye; *b′.* eye magnified; *b″.* frontal
eyes; *c.* the back; *d.* the tail; *e.* branchiæ.

Eyes in front of the shield. See V. I. p. 393. (Original.)

Fig. 2. View of the under surface of Fig. 1, shewing the crustaceous legs beneath the shield (*a*), and the swimming feet, bearing the Branchiæ (*e*), beneath the body (*c*). Scale, one seventh of nat. size.

Fig. 2. *e'*. Swimming feet, (see Fig. 2 *e*,) enlarged to the scale of Fig. 1.

Fig. 2. *e''*. Posterior surface of one of the swimming feet, bearing the fibres of the Branchiæ. (Original.)

Fig. 3. Front view of magnified figure of Branchipus stagnalis. 3. *b*. The left eye mounted on a peduncle. 3. *b'*. The right eye still more magnified. (Original.)

Fig. 4. Side view of Branchipus stagnalis, nat. size.

Fig. 5. Magnified view of the back of Branchipus stagnalis. See V. I. p. 394. (Original.)

Fig. 6. View of the back of a Serolis from Senegal, given by M. Dufresne to Dr. Leach. See V. I. p. 392. (Original.)

Fig. 7. View of the under surface of Fig. 6, shewing the union of crustaceous legs with the membranous branchiæ, *e*.* (Original.)

Fig. 8. Magnified view of the Branchiæ at Fig. 7, *e*.

Fig. 9. Back of Asaphus caudatus, from Dudley, in the collection of Mr. Stokes. (Original.)

Fig. 10. Side view of the left Eye of Fig. 9, magnified.

Fig. 10'. Another Eye of Asaphus caudatus, in the collection of Mr. Bright, from the W. side of Malvern Hill. In the front of this fossil are circular depressions on the stone, from which the petrified lenses have fallen out; on each side, the lenses remain in their natural place. (Original.)

* Figs. 3, 5, 6 and 7, are from original drawings by Mr. Curtis in the collection of Mr. C. Stokes.

Fig. 11. Anterior segment of the left Eye of Fig. 9, still more highly magnified, to shew the circular lenses set in their respective margins, each surrounded by six minute tubercles. (Original.)

Fig. 11'. Magnified view of a portion of the eye of Calymene macrophthalmus. (Hoeninghaus.)

Fig. 12. Under surface of the anterior portion of the shield of Asaphus platycephalus, from Lake Huron. An unique specimen, shewing at *f.* an entrance to the stomach, analogous to that in recent Crabs. . See Geol. Trans. N. S. Vol. i. Pl. 27. (Stokes.)

PLATE 46.   V. I. p. 389 et seq.

Figs. 1. 2. 3.   Calymene Blumenbachii, from the Transition Lime-stone of Dudley. *a.* The shield covering the head. *a'.* Lateral portion of the shield, separated by a suture from *a.*; the central part of this suture forms the lateral Margin, or Rim of the cavity of the Eye. This Margin is composed of two parts, united to receive the Lens, like the rims that enclose the edges of the glasses, in a pair of Spectacles. The Lens has usually fallen out from the Eyes of fossils of this species, as often happens after death in the Eyes of the recent Grapsus pictus, and also in the common Lobster. *b.* The Eye. · *c.* The dorsal portion, composed of articulating plates, that move on one another like the plates of a Lobster's tail. *d.* The tail.

Fig. 1. Side view of the Animal rolled up like an Oniscus. (Scharf.)

Fig. 2. View of the back of the Animal expanded for swimming; the Tail *d*, is composed of plates that had no moveable articulations. (Original.)

Fig. 3. Front view of the same Animal rolled up; the

shell, in this position, must have given perfect protection to the soft parts of the body enclosed within.   (Scharf.)

Fig. 4. Side view of Calymene macrophthalmus, rolled up, with its tail closed on its shield. (Curtis.)

Fig. 5. Front view of another specimen of C. Macroph-thalmus, rolled up like Fig. 4. The Eyes in fossils of this species are usually well preserved, and their facets large. (Curtis.)

Fig. 6. Asaphus tuberculatus; a highly ornamented species from the Transition lime-stone of Dudley; in the collection of Mr. Johnson, of Bristol. The back alone is composed of flexible plates. (Curtis.)

Fig. 7. Asaphus De Buchii, from the Transition slate of Llandilo; the tail is surrounded with an inflexible Margin, slightly fluted. (Brongniart.)

Fig. 8. Restoration of Paradoxoides Tessini, (Brong-niart. Hist. Nat. de Crustacés, Pl. IV. Fig. 1.)

Fig. 9. Ogygia Guettardii, (Brongniart, Hist. Nat. de Crustacés, Pl. III. Fig. 1.)

Fig. 10. Highly ornamented tail of Asaphus gemmu-liferus, (*Phillips*), from the Transition lime-stone of Dublin, magnified four times.   (Curtis.)

Fig. 11. Tail of Asaphus caudatus, from Carboniferous limestone, at Beadnell, Northumberland; in the collection of the Geol. Soc. of London. (Original).

Fig. 12. Tail of Asaphus caudatus, from Transition limestone, near Leominster; in the Oxford Museum.

PLATE 46'.   V. I. p. 406.

Fig. 1. Back of a fossil Scorpion of a new genus (Cy-clophthalmus) found by Count Sternberg in the Coal formation of Bohemia, in a quarry of sandy argillaceous Schist, sufficiently hard to be used for building. Nat. size. (V. I. p. 407, Note.)

Even the skin, hairs, and pores of the tracheæ of this animal are preserved.

In the same stone are many carbonized fragments of Vegetables, and on the right of the body is a large fossil Nut (a); this side of the animal has been laid open by cutting away the stone. (Sternberg.)

2. Lower surface of the same animal, discovered in splitting the stone in search of fossil Plants; nat. size. Near the point of the right claw, is a fragment of the tail of another and larger Scorpion. (See Pl. 46", Fig. 13.) We have here also the side of the same nut that is seen in Fig. 1. a. This trifid nut exhibits traces of the structure of the outer coating in which it was inclosed. (Sternberg.)

3. Magnified representation of the Head and Eyes. See V. I. p. 407. (Sternberg.)

4. Magnified jaw, armed with teeth, and partially covered with minute hairs. (Sternberg.)

5. Hairs on Fig. 4, highly magnified. (Sternberg.)

6. Magnified representation of a portion of the skin, consisting of two divisible layers. See V. I. p. 408. (Sternberg.)

7. Magnified impressions of muscular fibres connected with the legs. (Sternberg.)

PLATE 46". V. I. p. 409.

Fossil Insects, Arachnidans, and Limulus.

The following description of the Insects represented in this Plate is founded on information received from Mr. Curtis and Mr. Samouelle.

Figs. 1 and 2 belong to the family of Curculionidæ, of which the Diamond beetle is a familiar example. They were discovered by Mr. Wm. Anstice in nodules of Iron stone from the Coal formation of Coalbrook Dale.

Fig. 1 nearly resembles some of the South American types of Curculio, but the antennæ are longer and stronger than is usual in living species. Only the back of the head is visible, with faint indications of the place of the eyes; the Rostrum is not apparent, it probably descends into the Iron stone beneath, and this position will explain the appearance and place of the Antennæ.

The Elytra seem to have been connate towards their lower extremity, but their line of junction is visible towards the Thorax. The substance of the Elytra and Thorax, and of portions of the legs is replaced by white Iron ore, having the lustre of Satin.

Mr. Curtis conceives that the tufted appearance of the legs may have been caused by fungi formed after death, as often happens in tropical climates. The enlargement of the Femur of the hindmost leg in our fossil is a character peculiar to the Curculionidæ.* (Original.)

Fig. 2. Mr. Samouelle considers this extinct fossil species to approach most nearly to the Brachycerus apterus of Africa.† (Original.)

---

* Until more perfect data are found, on which generic characters can be established, I propose to designate this Insect by the provisional name of Curculioides Ansticii.

† The animal lies on its back with the left side raised upwards, and exhibiting a portion of the exterior surface of the left Elytron. At *a. b.* are the remains of antennæ, and near the base of *a,* ap-

Fig. 3. Limulus trilobitoides (*nobis*) forming the Nucleus of a nodule of Iron ore from Coalbrook Dale. V. I. p. 396.* (Original.)

parently a fragment of the proboscis; the legs are all imperfect; the thorax is very large, and only its inferior surface is visible, being exposed by the removal of the pectoral portion of the trunk; this surface is covered with irregular indentations, which represent the hollow interior of a series of spinous tubercles, and verrucose projections on the back of the thorax.

In the centre of the thorax is a compound depression larger than the rest, indicating the presence of a corresponding projection on the back.

Among living Curculionidæ irregular tubercles and projections of this kind occur on the thorax of the Brachycerus apterus.

The left Elytron only is distinctly visible, embracing with its margin the side of the Abdomen; its outer surface is irregularly and minutely punctate. Two spinous tubercles project from near its posterior extremity, and a corresponding tubercle from the extremity of the right elytron. Similar spines occur on the Elytrons of Brachycerus; and of some Curculionidæ of N. Holland. The abdominal rings are very distinct. I shall designate this Insect by the provisional name of Curculioides Prestvicii.

M. Audouin exhibited at the meeting of the Naturforscher at Bonn, in September, 1835, a beautiful wing of a neuropterous Insect, in a nodule of clay Iron stone, apparently also from the neighbourhood of Coalbrook Dale, which had been purchased at the sale of Parkinson's collection by Mr. Mantell, and transmitted by him to M. Brongniart. This wing is nearly three inches long, and closely resembles that of the living Corydalis of Carolina and Pensylvania; it is much broader and nearly of the length of the wing of a large Dragon Fly.

* Several specimens of this species are in the collection of Mr. Wm. Anstice at Madely Wood. Our figure is taken from a cast or impression of the back of the animal in Iron stone, in which the transverse lines across the abdominal segment are not very apparent; other specimens exhibit deep transverse flutings, externally resembling the separate segments of the back of a Trilobite, but apparently not dividing the shell into more than one abdominal Plate, nor admitting of flexure like the articulating segments of a Trilobite.

The transverse depressions on the back of the second segment of

Figs. 4—9. Elytra of Insects in the Oolitic slate of Stonesfield. Mr. Curtis considers all these to belong to the family Buprestis. (Original.)

Fig. 10. Leg of an Insect in the Stonesfield slate, Oxon, considered by Mr. Curtis to be that of a Curculio.* (Original.)

Fig. 11. A fossil Fly from the fresh water formation of Aix in Provence, in the collection of Mrs. Murchison. Mr. Curtis considers this Fly to be of the same species with one of those engraved in Fig. 11 of his Plate of Insects from this locality, in Jameson's Journal, Oct. 1829. (Original.)

Although it agrees with no living genus, he thinks it undoubtedly belongs to the family of Tipulidæ,

the body of this animal, form a character wherein it approaches nearer than the living Limulus to the structure of Trilobites. The articulation of the long awl-shaped tail with the body in Fig. 3, and in other specimens is very distinct. This Limulus is the Entomolithus monoculites of Martin, (*Petrifacta Derbiensia, Tab. 45, Fig. 4.*) and Belinurus bellulus of König, (*Icon. Sect.* Pl. XVIII. No. 230.) M. Parkinson, Org. Rem. iii. Pl. XVII. Fig. 18, has figured a similar fossil from Dudley, in iron stone of the Coal formation.

* Mr. Rr. C. Taylor mentions the occurrence of the wing covers of Beetles in the shale of the Danby Coal pits, in the Eastern Moorlands of Yorkshire. This shale has nearly the same place in the Oolitic series as the Stonesfield slate. See Loudon's Mag. Nat. Hist. V. iii. P. 361.

In the private collection of Dr. de Siebold at Leyden, I saw in Oct. 1835, a most beautiful and unique specimen of a Buprestis, from Japan, about an inch long, converted to Chalcedony. Even the antennæ and portions of the legs are distinctly preserved.

In the same collection are fragments of silicified trees, bored with tubular cavities, apparently by the larvæ of animals of this kind; and within these cavities, a quantity of dust, produced by the boring, was observed by M. Brongniart to be converted to Chalcedony. From this circumstance we may conjecture that the perfect insect was lodged in a similar tube, when it became transformed into Chalcedony. The surface of this Insect is covered with clusters of minute concentric rings of Chalcedony (Orbicules of Brongniart) so common in silicified fossil shells.

and is nearly related to the genus Bibio, which is now widely distributed, being common in Europe, and in N. and S. America.

> See Curtis Brit. Ent. Vol. iii. Pl. 138.

This fossil presents the under surface of the Animal.

Fig. 12. A fossil Spider from the *Miocene* Fresh-water formation at Aix, in Provence, in the collection of Mrs. Murchison; the under surface of the Animal is presented, and the little tubercles near the hinder part of the abdomen are Papillæ of the spinning organs, apparently, protruded by pressure.

> See Kirby and Spence, Introduction to Entomology, 4th edit. vol. i. p. 204; and Herold, von der Erzeugung der Spinnen im Eie, Tab. 11. Figs. 4. 9. 11. r. (Original.)

Fig. 13. From a drawing by M. Cotta of the fragment of a larger Scorpion, which is slightly delineated in Pl. 46′, Fig. 2, near the forceps of the smaller and more perfect Scorpion figured in that Plate. I received this drawing from Count Sternberg, in August, 1835. (Original.)

a. Dorsal scales of the abdomen.

b. Caudal segments.

c. Intestinal Canal. ?

d. Fragment of Intestinal Canal. ?

PLATE 47*.   V. I. p. 428, Note.

Fig. 1 and 2. Copied in part from the restoration of the Bradford, or Pear Encrinite (Apiocrirites rotundus) in Miller's Crinoidea, Pag. 19. Pl. 1. In Fig. 1. the arms are expanded, and in Fig. 2. nearly closed.

---

* Much value is added to this and the following Plates, relating to Crinoidea, by their having been engraved (except Pl. 48.) by a Naturalist so conversant with the subjects, as Mr. James Sowerby.

The length of the jointed flexible stems has been
taken from some entire stems in the collection of
Mr. Channing Pearce of Bradford, near Bath. Two
young individuals are attached to the calcareous
Pedicle or Base of the largest specimens. (Miller.)

Fig. 2. *a.* represents the remedial effect of calcareous
secretions in repairing an injury of the joints of the
stem. (Miller.)

Fig. 3. Pyriform Body of Apiocrinites rotundus, shewing at its upper extremity the internal disposition of
the bones surrounding the cavity of the stomach.
(Original.)

Fig. 4. Vertical section of another pyriform Body, shewing the cavity of the Stomach, and a series of lower
cavities, or hollow lenticular spaces, between the
central portions of the enlarged joints of the upper
portion of the vertebral column. Miller considers
these spaces as enlargements of the alimentary canal,
which descends through the axis of the entire column.

The surfaces of the joints of the vertebral column
are striated with rays, which articulate with corresponding rays on the adjacent Plates, and allow of
flexure without risk of dislocation; locking into one
another nearly in the same manner as those figured
in Pl. 49. Figs. 5. 7. 9. (Original.)

Fig. 5. Restored figure of Apiocrinites, 30-Dactylus,
copied from Miller's Crinoidea, Page, 96, Pl. 1.
Fig. 2. (See V. 1. p. 429. Note.)

B. Base and fibres of attachment.

D. Auxiliary side Arms*.

---

* These side arms afford a beautiful example of mechanical adap-
tations and compensations, which are thus described by Mr. Miller

Fig. 6. Body of Apiocrinites 30-dactylus (Nave Encrinite
of Parkinson) copied from Miller's Crinoidea, P. 98.
Pl. 11. (See V. 1. p. 429. Note.)

Q. Pectoral Plates.

R. Capital Plates.

X. Orifice of the Mouth, or Proboscis, capable of elon-
gation for sucking in food.

Fig. 7. Another Body of a Nave Encrinite, drawn by Mr.
J. Sowerby from a specimen in the British Museum.
The same is Figured by Parkinson, in his Organic
Remains Vol. II. Pl. XVII. Fig. 3. The lateral pro-
jections are the commencement of the side arms.
This specimen has been corroded with acid, and con-
sequently has lost the superficial Corrugations and

in his admirable Monograph on Crinoidea, p. 97. "The mechanism
of the joints of the side arms, where these insert into the column, is
well worthy of notice, particularly in old specimens. In the earlier
stage of their formation, the side arms being very short, and having
then little weight, a less firm mode of adhesion to the column than
becomes requisite at a subsequent period, being then sufficient, we
do not find more than one joint lodged in a socket, or concave impression on the column; but when increase of size renders a stronger
support necessary, two or three succeeding joints of the side arms
become imbedded in this socket, (for which its extension as already
noticed allows room) and these joints instead of being arranged in a
series branching off at right angles from the column, become oblique,
their direction inclining upwards, so as to aid in bearing the addi-
tional weight. The first joint of the side arms, where thus obliquely
inserted in the columnar socket, have that portion of their circumference which is presented towards the upper part of the column,
truncated, in such a curve as may fit them to the concavity of the
impression where they rest against it.

The surface of these joints, which fit into the columnar impression,
is smooth, being destined for adhesion only, but the articulating surface between the contiguous joints, where motion also is to be allowed,
exhibits the usual mechanism of radiated ridges and furrows. These
joints are convex on the side nearest the column, and concave on that
most remote."

Tubercles which appear on the surface of Fig. 6. (Original.)

X. Orifice of the Mouth.

PLATE 48. V. 1. p. 421.

Lily Encrinite, (Encrinites moniliformis,) from the Mus-chel-kalk, near Gottingen; in the Cabinet of the Marquis of Northampton. (Original.)

PLATE 49. V. I. p. 423. Note.

All the Figures in this Plate except Fig. 3, are taken from the Petrefacten of Dr. Goldfuss, Pl. LIII. and Pl. LIV. They are so fully explained in our Vol. I. p. 421 and Notes, as to supersede the necessity of any further detailed description.

Fig. 1. Restoration of the body and vertebral column of Encrinites moniliformis.

Fig. 2. Base of attachment.

Fig. 3. Portion of the summit of a vertebral column. (Original.)

Fig. 4. Longitudinal section of Fig. 3. magnified. (See V. I. p. 424. Note.)

Figs. 5, 7, 9. Joints from different parts of the vertebral column, showing the manner in which the articulating surfaces are crenulated to admit of flexure.

Figs. 6, 8, 10. Vertical sections through the axis of Figs. 5, 7, 9. shewing the forms of the internal cavity for the alimentary canal.

11—26. Profile and view of the articulating surfaces of joints, from various parts of the vertebral column. (See V. I. p. 425. Note.)

PLATE 50. V. I. p. 421, et seq.

Fig. 1. Fragment of the upper portion of Encrinites moniliformis, shewing the exterior of the Body, Arms, and Fingers, nearly closed around the tentacula. From a specimen belonging to Mr. Stokes. (Original.)

K. Arms.

M. Hand.

N. Fingers.

Fig. 2. Another fragment of the upper portion of the same species, reduced one-third, shewing the summit of the column, the exterior of the body, arms, and fingers, and the manner in which the Tentacula are folded when the animal is closed. See V. I. p. 427. Note. (Copied from Parkinson's Organic Remains, Vol. 2. Pl. XIV. Fig. 1.)

Fig. 3. Side View of one finger, with its tentacula, (Goldfuss, Pl. LIV.)

Fig. 4. Interior of the body. See V. I. p. 428. Note. (Miller, P. 40. Pl. II.)

Æ. Column.

E. Pelvis. ⁎⁄⁎ First Costal Plate. ⁎⁄⁎ Second Costal Plate.

H. Scapula.

Fig. 5. Articulating surface of the base. (Goldfuss, Pl. LIV.)

Fig. 6. Dissection of the Scapula. See V. I. p. 428, Note. (Miller.)

Fig. 7. Dissection of upper costal Plates. (Miller.)

Fig. 8. Dissection of lower costal Plates. (Miller.)

Fig. 9. Dissection of Pelvis. (Miller.)

Fig. 10. Summit of vertebral Column. (Miller.)

Figs. 11—18. Articulations of the Plates composing the

abdominal cavity. See V. I. p. 428. Note. (Copied
from Miller's Crinoidea, P. 41. Pl. III.

PLATE 51.   V. I. p. 434, and 439.

Fig. 1. Pentacrinites Briareus, (nat. size) on a slab of
Lias from Lyme Regis, covered with a large group
of the same animals, in the collection of the Geological Society of London. (Original.)

Fig. 2. Rare and beautiful specimen of Briarean Pentacrinite, from the Lias at Lyme Regis, in the collection of Mr. Johnson, of Bristol, shewing the plated
integument of the abdominal cavity, terminated upwards by a flexible Proboscis, and surrounded by
the commencement of the arms and fingers. This
part of the animal is very seldom preserved. See
V. I. p. 439. (Original.)

PLATE 52.   V. I. p. 432.

Fig. 1. Recent Pentacrinus Caput Medusæ, from the
bottom of the sea, near the I. Nevis, in the W. In-
dies, reduced from the Figure in Miller's Crinoidea,
P. 48, Pl. I.   In the front of this Figure, two of
the arms with their hands and fingers are much
smaller than the others, and shew that these animals, when mutilated, have the power of reproducing
lost parts.

D. Auxiliary side arms, articulating at distant intervals,
with the vertebral column; these also, when mutilated, are reproduced.

$\frac{a}{V.}$ First costal plate.

$\frac{b}{T.}$ Second costal plate.

H. Scapula.

I. Interscapulary joint.

Miller's description of this recent Type, of a family

of which a few individuals only have hitherto been found, affords examples of many very delicate and beautiful mechanical contrivances, which throw important light on corresponding parts of the fossil species of this, and of kindred genera that abound in strata of the Secondary series, and more especially in the Lias. (See V. I. pp. 432. 433. 436.)

Fig. 2. Pentacrinus Europæus, discovered in the Cove of Cork, and on other parts of the coasts of Ireland, by J. V. Thompson, esq. (See V. I. p. 432.) In this figure several Individuals in different stages of development, adhere by the base of an articulated column to the stem of a Coralline.

Fig. 2'. One of the Individuals magnified and fully expanded. See V. I. p. 433.

Mr. J. V. Thomson has more recently conjectured that the Pentacrinus Europæus, which in early life is fixed by its stem to other bodies, is produced from the ovum of the Comatula, and becomes afterwards detached, and forms a perfect Comatula, capable of moving freely in the Ocean; at one time crawling amongst sub-marine Plants, at others floating, or swimming like Medusæ. (See Proceedings of Royal Society, London, June, 1835.)

Fig. 3. Small Briarean Pentacrinite, adhering to a fragment of Jet from the Lias at Lyme Regis. (See V. I. p. 437, Note.)

Fig. 4. Fragment of the column of Pentacrinites subangularis. The Vertebræ are nicely articulated to admit of flexure without risk of dislocation. The uppermost joint *d*. shews the lateral cavities for the articulation of auxiliary side arms. (Goldfuss. Pl. LII. f. g.)

Fig. 5. Vertical Section of Fig. 4. In this Fig. and in

Fig 4, the joints are of three degrees of magnitude; those at *a.* being the largest, those at *c.* the smallest and thinnest, and those at *b.* of an intermediate size. The edges of *c.* appear at the surface only upon the salient portion of the column, Fig. 4. (See V. I. p. 436, Note.)

Figs. 6, 7, 8, 9, 12, 13. Portions of the vertebral column of Pentacrinites basaltiformis. 6, 8, 12, shew the stellated crenulations on the articulating facets of different parts of the column; 7, 9, shew the tubercles on the exterior of each columnar joint, for the attachment of cortical contractile fibres. 13. *d,* shews the articulating facets of the auxiliary side arms. (Goldfuss).

Fig. 10. Articulating facet of a columnar joint of Pentacrinites scalaris. (Goldfuss. Pl. LII. 3. h.)

Fig. 11. Fragment of a column of the same species. The joint *d.* bears sockets for the articulation of the side arms. The other joints have large tubercles for the attachment of cortical fibres. (Goldfuss, Pl. LII. 3. *p.*)

Figs. 14, 15, 16, 17. Articulating surfaces of joints in different parts of the column in Pentacrinites subangularis. The mechanism of each star seems differently disposed, to modify the amount of motion required at their respective places in the column. The tubercular surfaces between the rays or petals of the star indicate the action of the intervertebral contractile fibres. (Goldfuss, Pl. LII. 1. *m. n. o. p.*)

PLATE 53. V. I. p. 434, Note, et seq.

Fig. 1. 2. Upper parts of two nearly entire specimens of Briarean Pentacrinite, projecting in high relief from the surface of a slab, nearly two inches thick,

and entirely composed of a mass of petrified Ossicula of the same species of Pentacrinite. The surface of these fossils is covered with a delicate film
of Iron Pyrites, which gives them the appearance of
beautiful Bronze. (Original.)

1ª. Continuation of the stem of Fig. 1.

2ª. Portion of the stem of Fig. 2.

The length of these stems when entire, was three
or four times that of the fragments here remaining.

Upon the stem 2ª, nearly all the side arms retain
their places in the grooves on each side of the salient
angles of the pentagonal column; they diminish in
size as they approach its upper extremity. This is
also distinctly seen at the upper end of the column
of Fig. 1.

$\frac{*}{F.}$ First costal plate. $\frac{*}{F.}$ Second costal plate.

Fig. 3. Portion of a third column retaining nearly all its
auxiliary side arms in their natural place.

Fig. 3ª. Continuation of the same column deprived of
the side arms.

Fig. 4. Portion of another column, with traces of a few
side arms rising from the lateral grooves.

Fig. 4ª. Continuation of Fig. 4.

Fig. 5. Fragment of another column, the joints of which
are so much bent without dislocation, as almost to
give the column the appearance of a spiral disposition.

Fig. 6. Body of a Briarean Pentacrinite and summit of
its column, shewing the interior of the ossicula that
surround the abdominal cavity.

E. Pelvis. $\frac{*}{F.}$ First costal Plate. From a specimen in the
Oxford Museum. (Original.)

Fig. 7. Fragment of a column in the collection of Mr.
J. Sowerby, shewing the oblique articulation of the

base of the side arms, with the larger joints of the
vertebral column. See V. I. p. 439. Note. (Original.)

Fig. 8. Magnified Section of a portion of a Column in
the Oxford Museum. The joints, as in Pl. 52, Fig.
4, 5, and in Pl. 49, Figs. 3, 4, are alternately thicker
and thinner; with a third, and still thinner joint
interposed between them. See V. I. p. 435, Note.
(Original.)

Fig. 8ᵇ. Nat. size of Fig. 8.

Fig. 8ᶜ. Portion of a Column, shewing the manner in
which the edges of the thinnest plates, c, are visible
along the salient angles only. In the intermediate
grooves the thicker plates, of the first and second
sizes, a, b, overlap and conceal the edges of the
thinnest plates, c. The principle of this mechanism
is the same as in Pentacrinites subangularis, Pl.
52, Figs. 4, 5, and in Encrinites moniliformis, Pl.
49, Figs. 3, 4; but the circular form of the column
in the latter, causes the smallest plate, c, to be
visible around its entire circumference. See V. I.
p. 435, Note. (Original.)

The bases of two side arms are seen in two of the
grooves, articulating with the uppermost large joint
of this column. On other large joints are seen the
sockets from which similar side arms have fallen.

Figs. 9, 10, 11, 12, 13. Various stellated forms on the
articulating surfaces of Vertebræ, preserved in the
dislocated mass beneath Figs. 1, 2. These petal-
shaped, and crenated rays were probably adapted
to produce various degrees of flexibility, according
to their respective places in the column. The small
Vertebra on Fig. 13, is derived from another individual. (Original.)

The aperture at the centre of all these Vertebræ
was for the passage of the alimentary canal, which

Miller considers to have sent off ten branches at
every joint, five to the interior and five to the exterior of the petals.

Fig. 14. One of the largest auxiliary side arms. Some
of these contained more than 100 joints. See V. I.
p. 438. (Goldfuss.)

a, b, c. represent different forms of the joints at different
parts of the side arms, with their nicely adjusted
articulating surfaces.

Figs. 15, 16, a, b, &c. Various modifications of the
articulating surfaces of the joints composing the
fingers and tentacula. (Goldfuss, Pl. LI.)

Fig. 17. Magnified extremity of one of the tentacula.
The two last joints form a very delicate pair of pin-
cers, to lay hold on its prey. (Original.)

PLATE 54. V. I. p. 442.

Fig. 1. Caryophyllia arbuscula, nat. size, with the animals expanded. (Mem. du Mus. d'Hist. Nat. Tom.
6, Pl. 15, f. 2.)

Fig. 2. The animal of Fig. 1. magnified; as seen from
above.

Fig. 3. Vertical section of the cup of Meandrina laby-
rinthica, with the animal placed within it. (Mem.
du Mus. d'Hist. Nat. Tom. 6, Pl. 16, 10 b.)

Fig. 4. a. The common Actinia, or Sea Anemone, expanded. b. The same contracted within its external
skin. (Encyc. Method. Pl. 72. 6.*)

Fig. 5. Madrepora gyrosa. (Ellis. Zooph. Tab. 51,
Fig. 2.)

---

* This animal has no calcareous cell, but contracts itself into a
tough fleshy sac, see Fig. 4 b. At a. the Tentacula are represented
in a state of expansion. Some of these Polypes present the same
display of brilliant colours as many of those which construct persistent calcareous cells.

Fig. 6. Section of the animal of Meandrina viridis, and of the coral in which it is placed.

Fig. 7. Animals of Meandrina limosa as seen from above, and magnified; they are placed in confluent stellated cells similar to those in Fig. 5.

Fig. 8. One of the same, seen in profile, with the edges of its coralline plates behind the tentacula. (Mem. du Mus. d'Hist. Nat. Tom. 6, Pl. 15. 4.)

Fig. 9. Caryophyllia Smithii, from Torquay. Nat. size.

Fig. 10. The same, with its animal partially expanded, within the centre of the coral.

Fig. 11. The animal expanded, and seen from above. (Zoological Journal, Vol. 3. Pl. 13.)

PLATE 55. V. I. p. 466.

Fig. 1. A. B. C. Trunk, and dichotomous branches of a fossil tree, Lepidodendron Sternbergii, found in the roof of a coal mine at Swina, in Bohemia. (Sternberg, Tab. I.)

Fig. 2. The extremity of a branch with leaves attached to it, from ten to twelve inches long.* (Sternberg, Tab. II.)

Fig. 3. Extremity of another branch, with indications of fructification somewhat resembling a cone. (Sternberg.)

PLATE 56. V. I. p. 469, et seq.

Extinct Plants from the Coal Formation.

Fig. 1. Copied from a sketch by Mr. Sopwith, of the base of a large trunk of Sigillaria standing in 1803, in the cliff at Bog Hall, near Newbiggin, on the

---

* By an error in copying this figure the branches are made too broad in proportion to the leaves.

coast of Northumberland. This fragment is about
five feet high, and two feet three inches in diameter
at its base.* Scale one-twenty-fourth. (Sopwith.)

2. Fragment of the bark on the trunk of a Sigillaria,
from Earl Fitzwilliam's coal mine at Elsikar, near
Rotherham. In this mine many large trunks are
seen inclined in all directions, and some nearly vertical. (See V. I. p. 470, Note.) The bark is converted
into a thin lamina of coal, and remains attached to
the lower portion of this specimen. It exhibits on
its outer surface scars formed by the articulations
of the bases of leaves; these are penetrated near
their centre by three apertures for vessels that
passed from each leaf into the trunk. The decorti-
cated upper part of this specimen presents an impression of its striated internal surface, and exhibits
beneath each scale two oblong parallel apertures,
through which the vessels from a leaf penetrated
the trunk. Scale one-half. (Original.)

The substance of the trunk must have been in
a state of decay, before the mud, which is now har-
dened into shale, could have entered the interior of
the bark. When trunks of this kind are inclined
at an angle exceeding 45°, they are usually distended with sandstone, or sandy shale; when at a
less angle than 45°, they are most commonly compressed, and have only a thin flat portion of shale,
formed of indurated mud within their bark. The
bark, wherever it has not perished, is converted to
coal.

2'. Articulating leaf-scar on the exterior of the bark of
another large trunk of Sigillaria from Elsecar. Nat.

* M. Ad. Brongniart found a stem of Sigillaria in a coal mine at
Essen in Westphalia, which was dichotomous near its top.

size.　On comparing this scar with those upon the
bark of Fig. 2, it may be seen that the different
modes of articulation of the leaves with the cortical
integument present obvious characters, on which
specific distinctions may perhaps most easily be
established, in this very obscure and curious family
of extinct plants.　See various figures of these leaf-
scars in Lindley and Hutton's Fossil Flora, Plates
55. 56. 57. 71. 72. &c.　In Figs. 2, and 2', as in
many other species, decurrent lines are visible on
both sides of the scar.　(Original.)

Fig. 3. Ulodendron Allanii, (*nobis*) scale one-fifth.
See V. I. p. 475. Note.　Drawn from a plaster cast
of an impression on sandstone, in the Museum of
the Royal Society of Edinburgh from the Coal formation at Craigleith.　This sandstone has formed
a natural mould on the outer surface of a stem,
which has entirely perished; our cast gives a fac-
simile of the small rhomboidal scales, and of three
large round scars on the exterior of the trunk.
This impression has been figured, in an inverted
position, by Mr. Allan in Vol. IX. Trans. Royal
Soc. Edin. 1823. Pl. XIV. p. 236.　(Original.)

Our figure represents the trunk in its natural
position.　In the centre of each scar is a cavity,
indicating the place of attachment of a cone.　The
upper portion of each scar is marked with furrows,
produced by pressure of the long radiating scales
at the bottom of the cone.　This pressure has
nearly obliterated the smaller rhomboidal scales of
the bark, in those parts where the furrows are
deepest; on the lower portion of the scars, the
scales of the bark have been but slightly modified
by pressure of the cone.

Fig. 4. A single scar formed by the attachment of a cone
of another species, Ulodendron Lucasii, (*nobis*) discovered by Mr. Lucas in the S. Wales Coal field
near Swansea. Some scales and speared-shaped
leaves of the trunk are still preserved around the margin of this scar. As the bark has fallen off, we have
only the impression of its inner surface. This surface exhibits small apertures, through which vessels
entered from beneath the bark-scales into the trunk.
On the upper part of the disk, the traces of many
of these vessels have been obliterated by pressure
of the cone. Scale one-fourth. (Original.)

Fig. 5. Ulodendron Stokesii. (*nobis*) A large oval scar,
(4¼ inches in its longer, and 3½ inches in its shorter
diameter) preserved in shale from an unknown locality in the English Coal Formation. On the
margin of this scar are the remains of rhomboidal
scales, and impressions of scales, and a few small
leaves. Within the disk a few fragments only of
the bark remain near its upper margin. Near its
centre, is the mark of the insertion of the stem of
a large cone. The lower half exhibits a series of
small tubular cavities, marking the place of vessels
which passed from the bark into the trunk, one
beneath each of the bark-scales that have fallen off.
In the upper half of the Scar, there are but slight
traces of these cavities, and the surface is marked
with furrows, produced by pressure of the long
radiating scales of the base of the cone. Scale one-
fifth. (Original.)

Fig. 6. Ulodendron Rhodii. (*nobis*) Scar on a scaly
stem, from the Coal field of Silesia, figured by Rhode
in his *Beitrage zur Pflanzenkunde der Vorwelt*, L.
2. Pl. 3. Fig. 1. The lower portion of this Scar

retains the bark-scales modified by pressure of the
Strobilus or cone that grew from the centre of the
disk. The upper portion of the Scar is without indications of bark-scales, and is covered with radiating furrows, impressed on it by the long slender
scales of the base of the Strobilus, which have obliterated the bark-scales.*

The character of this scar approaches to that of
Fig. 5, but its proportions differ, measuring 3¼
inches in the longer, and 2½ inches in the shorter
diameter. The scaly bark (which in Fig. 5 has
been almost entirely removed from the area of the
scar), is preserved on the lower portion of the disk
of Fig. 6. Scale two-ninths. (Original.)

Fig. 6'. Cast of Ulodendron Conybearii (*nobis*) formed
by Pennant sandstone of the Coal formation at
Stapleton near Bristol. This cast expresses the
exact form of an oval scar, or cavity on a stem
from which a cone had fallen off.

The disk is covered with slight ridges and furrows,
radiating in all directions from the point of insertion of the cone, and formed by pressure of its
lowest scales upon the portion of the stem to which
it was attached. Beneath the point of insertion, a

---

* The portions above and below the line drawn across Fig. 6, are
copied from two scars in Rhode's figure. Rhode considers these
impressions to be flowers, and the compressed bark-scales to be the
Petioles of the flower, and has represented the trunk in an inverted
position.

As, in every species of Ulodendron which we have seen, the furrows produced by scales at the base of the cone, are deepest on the
upper portion of the Scar, we infer from this circumstance that the
cones were inclined upwards and inwards, with their axis approxi-
mating to that of the stem from which they issued.

few small scales of the bark remain adhering to the Sandstone. Scale one-fourth. (Original.)

Fig. 7. Portion of the Trunk of Favularia, one-fourth nat. size. This plant is distinguished by the tessellated appearance of the scales, which cover the space between each fluting of the Bark. In the centre of the area of each scale is a club-shaped scar, which gave origin to a leaf; it was a dicotyledonous plant, probably allied to Sigillaria; and its stem must have been covered with a mass of densely imbricated foliage. In the Genus Sigillaria the leaves were more distant from one another. The Rows of scars are separated by a groove, Fig. 7. *b.*; their disposition in the vertical direction is indicated by the line *a*. (Lindley, Foss. Fl. Pl. 73.)

Fig. 8. Reduced from Lindley and Hutton's figure (Pl. 31) of the central portion of a Stigmaria ficoides, from Shale in the roof of the Jarrow colliery near Newcastle. We have here a view of the inferior surface of this curious plant. Its dome-shaped hollow central trunk, or stem, was three feet in diameter, and fitted to sustain horizontally in a floating position the numerous long branches by which it was surrounded; these divide into two, at a certain distance from the Trunk. When perfect, and floating in water, its appearance must have resembled the form of an Asterias. On the two longest branches, *a. b.* is seen the longitudinal depression, which is usually adjacent to the small internal woody axis of these branches, and from its position in this fossil, we learn that the place of this depression was on the inferior surface of each branch. Scale one-twenty-fourth. (See V. I. p. 476.)

Fig. 9. Vertical section of the dome-shaped trunk of Stigmaria, shewing the relative position of the branches. (Lindley and Hutton.)

Fig. 10. Restored portion of a branch of Stigmaria, shewing the manner in which the long cylindrical leaves proceeded from the tubercles around its surface to the length of many feet. In front, extending from *a.* to *b.* is seen the depression adjacent to the internal eccentric woody axis *a.* From *b.* to *c.* this axis is laid bare by the removal of a portion of the sandstone. This part of the axis is drawn from a specimen in the Oxford Museum. Scale one-seventh. (Original.)

Fig. 11. Fragment of a branch of Stigmaria, shewing the character of the Tubercles, which formed articulations with the bases of the leaves. The enlargement of the leaf towards its base (*a*) seems to have been calculated to strengthen this part, and to afford space for the articulating socket. This socket formed, with the spherical tubercle, an universal ball and socket joint, admitting of motion in every direction to a long cylindrical leaf floating in water. Scale one-half. (Sternberg.)

PLATE 56ᵃ.   V. I. p. 483 et seq.

Appearances presented by longitudinal and transverse sections of recent and fossil Coniferous woods, cut into thin slices, and magnified 400 times. (Nicol.)

Fig. 1. Longitudinal Section of Pinus Strobus, cut parallel to a medullary ray.

Fig. 2. Transverse Section of the same.

a. a. Portions of concentric annual layers.

Fig. 3. Longitudinal Section of Araucaria Cunninghami.

Fig. 4. Transverse Section of the same.

Fig. 5. Longitudinal Sections of Araucaria excelsa, shewing polygonal disks, in double and triple rows, on the surface of the longitudinal tubes. Some of the tubes are without disks, as in all Coniferæ.

Fig. 6. Transverse Section of Araucaria excelsa.

a. Portion of concentric annual layer.

Fig. 7. Radiating and concentric structure of a branch of Pinus, as seen by the naked eye in a transverse section; the microscopic reticulations are omitted. (See V. I. p. 486. Note.)

a. a. Concentric annual layers, indicating periodical growth.

Fig. 8. Longitudinal Section of Pinus, shewing the relative positions of the longitudinal vessels and medullary rays.

a, Longitudinal vessels, forming the woody fibres.

b, Medullary rays.

PLATE 57.　V. I. p. 494.

Sections exhibiting the silicified remains of Coniferæ and Cycadeæ, in their native bed, between the Portland and Purbeck stone, on the coast of Dorsetshire.

Fig. 1. Appearance of trunks and roots of large Coniferous trees, and of trunks of Cycadites, in the black earth, which formed the soil of an ancient Forest in the Isle of Portland. (De la Beche.)

Fig. 2. Remarkable concentric Ridges of Stone, around the erect stump of a Fossil Tree in the Isle of Portland. See V. I. p. 495. Note. (Henslow.)

Fig. 3. Inclined position of the petrified stumps of large Coniferæ, and of the bed of black mould and pebbles in which they grew, near Lulworth Cove, on the Coast of Dorset. (Buckland.)

PLATE 58. V. I. p. 493.

Cycas revoluta, producing Buds from the axillæ of the scales, or persistent bases of leaves, that form the false bark. Drawn from a plant in the conservatory of Lord Grenville at Dropmore, 1832.

PLATE 59. V. I. p. 494.

Fig. 1. Zamia pungens, with its fruit, as it grew at Walton on Thames, 1832, in the Conservatory of Lady Tankerville. (Lambert.)

Fig. 2. Transverse section of the trunk of Zamia horrida, from the Cape of Good Hope. (Buckland.)·

Fig. 3. Transverse section of a young trunk of Cycas revoluta. See Geol. Trans. Lond. 1828. N. S. Vol. ii. Pt. 3. Pl. 46. (Buckland.)

PLATE 60. V. I. p. 497.

Fig. 1. Silicified trunk of Cycadites megalophyllus, from the Dirt bed in the Isle of Portland. (Original.)

Fig. 2. Portion of the Base of Fig. 1. See V. I. p. 497, Note. (Original.)

In Plates 60, 61, A represents the central mass of cellular tissue. B the single circle of radiating woody plates. C the circle of cellular tissue, surrounding B. And D the case or false Bark, surrounding C. And in Pl. 61, Fig. 1, *b*, represents a second circle of radiating woody plates.

PLATE 61. V. I. p. 497. Note.

Fig. 1. Silicified trunk of Cycadites microphyllus, from the Isle of Portland, with numerous buds rising from the axillæ of the Petioles. (Original.)

Figs. 2, 3. Vertical sections of agatised Petioles, composing the false bark on the trunk of Cycadites
microphyllus, and of embryo Buds. In the bud,
Fig. 2. *d.* the division between the two woody circles
is not distinct. In Fig. 3. '*d.* it is very obvious; but
the intermediate circle of cellular tissue is represented only by a fine line. See V. I. p. 498, Note.
and p. 500, Note. (Original.)

In the sections of Pl. 61, Figs. 2, 3, and Pl. 62, the
following letters are used to indicate the same parts. *a,*
cotton, or down; *b,* integument of petioles or scales; *c,*
bundles of vessels; *d,* woody circles; *e,* imperfect woody
circles; *f,* cellular tissue; *g,* embryo bud; *h,* gum vessels.*

PLATE 62. V. I. p. 498, Note.

Fig. 1. Longitudinal section of a Petiole of Zamia spi-
ralis magnified two times. It exhibits four bundles
of vessels passing longitudinally through the cellular
tissue, which is interspersed with gum vessels. V.
I. p. 499. (Original.)

A. Transverse section of Fig. 1. magnified, and showing
the irregular disposition of the bundles of vessels.
(Original.)

c'. Magnified view of one of the bundles of vessels at
A, *c.* (Original.)

B. c". Magnified transverse section of a bundle of vessels in the petiole of Zamia horrida. (Original.)

Fig. 2. Longitudinal section of a portion of an agatised
petiole of Cycadites microphyllus, from Portland,
magnified four times. The down or cotton at *a,* is

---

* These very beautiful and instructive sections were presented to
me by Mr. Witham, being portions of a trunk which I had placed
at his disposition.

most beautifully preserved, and the integuments of
the petiole *b*, longitudinal vessels *e*, and gum vessels *f*, correspond with those in Fig. 1. See V. I.
p. 499, Note. (Original.)*

Fig. 3. Transverse section of a portion of the lowest
Petioles, in Pl. 61. Fig. 3, *b*, *c*, magnified four
times. The disposition of the bundles of vessels is
nearly parallel to the integument of the Petiole.†

d. Magnified portion of the double woody circle, within
the Embryo bud, Pl. 61. Fig. 3, '*d*.

d'. More highly magnified portion of the embryo double
woody circle *d*.

c'. More highly magnified section of one of the bundles
of vessels, adjacent to *c*.

These bundles of vessels exhibit, in their transverse
Section, a series of minute tubes, arranged in rows, and
between these rows, opaque plates of compressed cellular
tissue, resembling portions of medullary rays.

The fibrous structure of the integument is preserved in
several parts of *b*. See V. I. p. 501, Note. (Original.)

---

* Mr. Robert Brown has noticed in the cellular tissue of a silicified
trunk of Cycadites, portions of Chalcedony bearing the form of ex-
travasated gum within the trunks of recent Cyadeæ. He has also
recognised spiral vessels, in the laminated woody circle of a mature
trunk of fossil Cycadites, and also in the laminated circle within a
silicified bud of the same, near its origin.

† A familiar example of a nearly similar disposition of bundles
of vessels, passing into the Petiole or leaf-stalk, may be seen in the
base of the fresh fallen leaves from a horse-chestnut tree, or in the
scars on a cabbage-stalk, from which leaves have fallen off.

PLATE 63.*   V. I. p. 503.

Fig. 1. Recent Pandanus, of S. America, twenty feet
high, with its fruit attached. (Mirbel.)

Fig. 2. Fossil fruit of Podocarya, from the Inferior Oolite,
near Charmouth, Dorset. Great part of the surface
is covered with a stellated Epicarpium; the points
of many seeds project in those parts (e) where the
Epicarpium is wanting. (Original.)

Fig. 3. Reverse of Fig. 2. shewing the seeds placed in
single cells (b) around the circumference of the fruit.
These seeds stand on a congeries of foot-stalks (d)
composed of long fibres, which terminate in the receptacle.(r) The surface of the receptacle is studded
over with small disks, in which these foot-stalks
originate. (Original.)

Fig. 4. Base of the same fruit, shewing the transverse
section of the receptacle (r), and the summits of
many abortive cells on the left side of the receptacle.
(Original.)

Fig. 5. A single seed of Podocarya converted to carbonate of lime. Nat. size. (Original.)

Fig. 6. The same magnified. (Original.)

Fig. 7. Transverse section of a seed magnified. Two
lunate marks, of a darker colour than the other part,
appear near its centre, f. See Fig. 8. and the
middle of Fig. 10. (Original.)

Fig. 8. Magnified portion of Fig. 3: shewing a withered

*   *Explanation of Letters of Reference.*

a. Stellated tubercles, each one covering the apex of a single seed.
b. Sections of the seed cells.
c. Bases of cells from which seeds have fallen.
d. Fibrous foot-stalks between the seeds and receptacle.
e. Apices of seeds uncovered.
f. Transverse section of seeds.

stigma in the centre of each hexagonal tubercle (*a*);
beneath these tubercles is a longitudinal Section of
the single cells (*b*), each containing one seed (*f*);
and in front of these cells are the hollow bases of
other cells (*c, c*) from which seeds have been removed.   (Original.)

Fig. 9. Another magnified portion, shewing the apices
of many seeds (*e*) from which the Epicarpium has
been removed.   (Original.)

Fig. 10. Another magnified portion, shewing at *a, b, c,*
more distinctly the same parts as at Fig. 8; and at
d, the upper portion of the fibrous foot-stalks beneath the bases of the cells, *c.*   (Original.)

Fig. 11. Summit of one of the drupes or groups of cells
into which the fruit of the recent Pandanus is
divided; shewing an hexagonal disposition of the
coronary tubercles, each bearing at its centre the
remains of a stigma, as in the Podocarya. See Figs.
16. 17.   (Original.)

Fig. 12. Exterior of a single seed-cell of Pandanus odoratissimus.   (Jaquin. Frag. Bot. Pl. 14.)

Fig. 13. Section of a Drupe of Pandanus odoratissimus.
The central cell containing a seed, is placed between
two abortive cells.   At the apex of each cell in this
drupe (*a*) is a withered stigma. (Roxborough Coro-
mandel. Pl. 96.)

Figs. 14, 15. Sections of a Drupe of Pandanus odoratissimus, shewing the seeds within the prolific cells
surrounded by a hard nut.   Beneath this nut is a
mass of rigid fibres like those beneath the seeds of
Podocarya.   (Jaquin.)

Fig. 16. Summit of the hexagonal tubercle at the apex
of a cell of Pandanus humilis, with a withered
stigma in the centre. (Jaquin. Frag. Bot. Pl. 14.)

Fig. 17. Side view of another tubercle of the same
species. (Jaquin. Frag. Bot. Pl. 14.)

PLATE 64.   V. I. p. 517.

Fig. 1. Fossil leaf of a Flabelliform Palm from the Gypsum of Aix in Provence. (Brongniart.)

Fig. 2. Upper portion of the Fossil trunk of a tree
allied to Palms (nearly four feet in diameter), from
the Calcaire Grossier at Vaillet, near Soissons, preserved in the Museum d'Hist. Nat. at Paris. See
p. 516, Note. (Brongniart.)

PLATE 65.   V. I. p. 529.

Fig. 1. Section across the Wednesbury Coal basin from
Dudley to Walsall. (Jukes.)

The extensive Iron foundries which cover the
surface of this district, and the greater part of the
manufactures in the adjacent town of Birmingham,
originate in the Coal and Iron ore, with which the
strata of shale in this Coal basin are richly loaded.

The Dudley Limestone, here found immediately
below the Coal formation, occurs usually at a much
greater depth in the series. The Mountain Limestone, Old red Sandstone, and Ludlow rocks, are
here wanting. (See Pl. 66, Fig. 1.)

Fig. 2. Section, shewing the basin-shaped disposition of
the Carboniferous strata in S. Wales. (Rev. W.
D. Conybeare.)

The richest beds of Coal and Iron ore are placed
almost immediately above the Mountain limestone.
(See pp. 65, 529.) It is to this district that our
Posterity must look for their future supply of Coals,
and transfer the site of their Manufactures, when

the Coal fields of the northern and central parts
of England shall be exhausted.*

Fig. 3. Section of inclined Carboniferous strata, over-
laid unconformably by horizontal strata of New
Red Sandstone, Lias, and Oolite, in Somersetshire.

This Section illustrates the manner in which Carboniferous strata have been elevated at their extre-
mities around the circumference of a basin, and
depressed towards its centre, and also intersected
by fractures or Faults. See V. I. pp. 527, 542.

In Section 1, 2, of this Plate, no notice is taken
of the Faults which intersect the Coal basins.

PLATE 66. V. I. p. 527, Note.

Fig. 1. Section of the strata composing the *Silurian
System*, and the lower part of the Carboniferous
System, on the frontiers of England and Wales.
(Murchison.)

Fig. 2. Appearance of Faults intersecting the Coal formation near Newcastle-on-Tyne, copied from a
portion of one of Mr. Buddle's important sections
of the Newcastle Coal field, in the Transactions of
the Nat. Hist. Society of Northumberland, V. I.
Pt. 3, Pl. XXI. XXII. XXIII.† The advantages

---

* The lower and richest beds of this Coal district are not only
raised to the surface, and rendered easily accessible around the external margin of the basin, but are also brought within reach in
consequence of another important elevation, along an anticlinal line,
running nearly E. and W. through a considerable portion of the
interior of the basin, in the direction of its longer diameter.

† I feel it a public duty to make known an act of Mr. Buddle,
which will entitle him to the gratitude of posterity, and has set an
example, which, if generally followed in all extensive collieries, will
save the lives of thousands of unfortunate miners, that must otherwise perish for want of information which can, at this time, be easily
recorded for their preservation. This eminent Engineer and Coal
Viewer has presented to the Natural History Society of Newcastle,

resulting from these Interruptions of the continuity of the strata are pointed out in pp. 543, 544.

A large portion of the surface of these strata near Newcastle is covered with a thick bed of diluvial Clay interspersed with Pebbles, in the manner represented at the top of this Section. The effect of this Clay must be to exclude much rainwater that would have percolated downwards into the Coal mines, had strata of porous Sandstone formed the actual surface.

PLATE 67. V. I. p. 559.

Fig. 1. represents the case of a valley of Denudation in stratified rocks, terminated abruptly by a cliff on

---

copies of his most important plans and sections, accompanied by written documents, of the under ground workings in the Collieries near that town, in which all those spaces are carefully noted, from whence the Coal has been extracted. Every practical Miner is too well acquainted with the danger of approaching ancient workings in consequence of the accumulation of water in those parts from which Coal has been removed. The sudden irruption of this water into a mine adjacent to such reservoirs is occasionally attended with most calamitous and fatal results. See History of Fossil Fuel, the Collieries and Coal Trade, 1835. P. 249 et seq.

The distates of humanity which prompt us to aid in the preservation of human life, no less than the economical view of rendering available at a future time the residuary portions of our beds of Coal, which will not now repay the cost of extracting them, should induce all proprietors and other persons connected with Coal Mines, and especially Engineers and Coal Viewers, to leave to their successors a legacy, which will to them be precious, by preserving minute and exact records of the state of the coal in their respective districts. It can, however, scarcely be expected, that such measures will be generally and systematically adopted throughout the many Coal fields of this country, unless the subject be legislatively taken up by those official persons, whom it behoves, as guardians of the future welfare of the nation, to institute due measures, whilst the opportunities exist, for preventing that loss of life and property, which a little attention bestowed in season, will preserve to posterity.

the sea-shore; this figure is intended to illustrate
two causes of the production of Springs by *descent*
of water from porous strata at higher levels; the
first, producing discharges in vallies of Denudation,
along the line of junction of porous with impermeable strata; the other, by the interruption offered
to descent of water by Faults that intersect the
strata.

The Hills A, C, are supposed to be formed of a
permeable stratum a, a', a'', resting on an impermeable bed of Clay b, b', b''.   Between these two
Hills is a Valley of Denudation, B.   Towards the
head of this Valley the junction of the permeable
stratum a, a', with the Clay bed. b, b', produces a
spring at the point S.; here the intersection of
these strata by the denudation of the valley affords
a perennial issue to the Rain water, which falls
upon the adjacent upland plain, and percolating
downwards to the bottom of the porous stratum a, a',
accumulates therein until it is discharged by numerous springs, in positions similar to S, near the head
and along the sides of the vallies which intersect
the junction of the stratum a, a', with the stratum
b, b'.   See V. I. p. 559.*

The Hill C, represents the case of a spring produced by a Fault, H.   The Rain that falls upon
this Hill between H, and D, descends through the
porous stratum a'', to the subjacent bed of Clay b''.

---

* The term *Combe*, so common in the names of upland Villages,
is usually applied to that unwatered portion of a valley, which forms
its continuation beyond, and above the most elevated spring that issues
into it; at this point, or spring head, the valley ends, and the *Combe*
begins.   The conveniences of water and shelter which these spring-
heads afford, have usually fixed the site of the highest villages that
are planted around the margin of elevated plains.

The inclination of this bed directs its course towards the Fault H, where its progress is intercepted
by the dislocated edge of the Clay bed *b′*, and a
spring is formed at the point f.   Springs originating
in causes of this kind are of very frequent occurrence, and are easily recognized in cliffs upon the
sea-shore.†   In inland districts, the fractures which
cause these springs are usually less apparent, and
the issues of water often give to the Geologist notice
of Faults, of which the form of the surface affords
no visible indication.   See V. I. p. 560, Note.

Fig. 2. Section of the valley of Pyrmont in Westphalia.
A cold chalybeate water rises in this valley at *d*,
through broken fragments of New Red Sandstone,
filling a fracture which forms the Axis of Elevation
of the valley.   The strata are elevated unequally
on opposite sides of this fracture.   See V. I. p. 561.
(Hoffmann.)

Explanation of Letters referred to in this Figure.

a. Keuper.

b. Muschel-kalk or shelly Limestone.

c. Variegated Sandstone.

d. Cold chalybeate Springs rising through a fracture on
the Axis of Elevation of the Valley.

M. The Muhlberg, 1107 feet above the sea.

B. The Bomberg, 1136 feet above the sea.

P. Pyrmont, 250 feet above the sea.

Fig. 3. Section reduced from Thomas's survey of the
mining district of Cornwall (1819) ; it exhibits

† Three such cases may be seen on the banks of the Severn near
Bristol, in small faults that traverse the low cliff of Red Marl and
Lias on the N. E. of the Aust Passage.   See Geol. Trans. N. S.
Vol. I. Pt. II. Pl. 37.

the manner in which the Granite and Slate near
Redruth are intersected by metalliferous Veins,
terminated abruptly at the surface, and descending
to an unknown depth; these Veins are usually most
productive near the junction of the Granite with the
Slate, and where one Vein intersects another. The
mean direction of the greatest number of them is
nearly from E. N. E. to W. S. W. They are intersected nearly at right angles by other and less numerous Veins called Cross Courses, the contents
of which usually differ from those of the E. and
W. veins, and are seldom metalliferous.

The Granite and Killas and other rocks which
intersect them, e. g. Dykes and intruded masses of
more recent Granite, and of various kinds of por-
phyritic rocks called Elvans (see Pl. 1, a 9. b. c.)
are considered to have occupied their present relative positions, before the origin of the fissures,
which form the metalliferous Veins, that intersect
them all. (See V. I. p. 550.)*

_____

* In Vol. I. P. 552, Note, a reference is made to some important
observations by Mr. R. W. Fox on the Electro-magnetic actions
which are now going on in the mines of Cornwall, as being likely to
throw important light on the manner in which the ores have been introduced to metallic veins.

The following observations by the same gentleman in a recent
communication to the Geological Society of London, (April, 1836,)
appear to contain the rudiments of a Theory, which, when maturely
developed, promises to offer a solution of this difficult and complex
Problem.

" If it be admitted that fissures may have been produced by changes
in the temperature of the earth, there can be little difficulty in also
admitting that electricity may have powerfully influenced the existing
arrangement of the contents of mineral veins. How are we otherwise to account for the relative positions of veins of different kinds
with respect to each other, and likewise of their contents in reference
to the rocks which they traverse, and many other phenomena ob-

PLATE 68. V. I. p. 563.

Section shewing the basin-shaped disposition of Strata belonging to the Tertiary and Cretaceous Formations, in the Basin of London, and illustrating the causes of the rise of water in Artesian Wells. See V. I. p. 564. Note. (Original.)

servable in them? Copper, Tin, Iron, and Zinc, in combination with the sulphuric and muriatic acids, being very soluble in water, are, in this state, capable of conducting voltaic electricity; so, if by means of infiltration, or any other process, we suppose the water to have been impregnated with any of these metallic salts, the rocks containing different salts would undoubtedly become in different or opposite electrical conditions; and hence, if there were no other cause, electric currents would be generated, and be readily transmitted through the fissures containing water with salts in solution; and decompositions of the salts and a transference of their elements, in some cases, to great distances, would be the natural result. But, on the known principles of *Electro-magnetism*, it is evident that such currents would be more or less influenced in their direction and intensity by the magnetism of the earth. They cannot, for instance, pass from N. to S. or from S. to N. so easily as from E. to W. but more so than from W. to E. The terrestrial magnetism would therefore tend, in a greater or less degree, to direct the voltaic currents through those fissures which might approximate to an east and west bearing, and in separating the saline constituents, would deposit the metal within or near the electro-negative rock, and the acid would be determined towards the electro-positive rock, and probably enter new combinations. Or, the sulphuric acid might, by means of the same agency, be resolved into its elements; in which case the sulphur would take the direction of the metal, and the oxygen of the acid, and in this way, the metallic sulphurets may have probably their origin; for, if I mistake not, the metallic *sulphates*, supposing them to have been the prevailing salts, as at present, would be fully adequate to supply all the sulphur required by the same metals to form sulphurets; indeed more than sufficient, if we deduct the oxyde of tin, and other metalliferous oxydes found in our mines. The continued circulation of the waters would, in time, bring most of the soluble salts under the influence of these currents, till the metals were in great measure separated from their solvents,

PLATE 69.   V. I. p. 565.

Fig. 1. Theoretical section, illustrating the Hydraulic
conditions of strata disposed in the form of Basins.
See Vol. I. p. 565, Note. (Original.)

Fig. 2. Theoretical section, shewing the effect of Faults
and Dykes on water percolating inclined and permeable Strata. See Vol. I. p. 566, Note. (Original.)

Fig. 3. Double Artesian Fountain at St. Ouen, near
Paris, raising water to supply a Canal basin, from
two strata at different depths. The water from
the lowest stratum rises to the greatest height.
See V. I. p. 562. Note. (Hericart de Thury.)

and deposited in the East and West veins, and near the rocks to
which they were determined by the electric currents."

In a Letter to the Author upon this subject (June 29, 1836), Mr.
Fox further remarks.

"It should be observed that in proportion as the deposition of the
metals proceeded, the voltaic action must necessarily have been
considerably augmented, so as to render it highly probable that the
metals were chiefly deposited at rather an early period in the history
of the containing veins ; and their intersection by other veins seems
to strengthen this probability."

Mr. Fox has found by experiment that when a solution of muriate
of Tin is placed in the voltaic current, a portion of the metal is determined towards the negative pole, whilst another portion in the
state of an oxide passes to the positive pole. This fact appears to
him to afford a striking illustration of the manner in which Tin and
Copper have been separated from each other in the same vein, or in
contiguous veins, whilst these metals also very commonly occur together in the same vein.

INDEX.

ACRODUS, a genus of fossil sharks, i. 288.

Actinocrinites, 30-dactylus, Miller's restoration of, i. 429.

Adapis, character and place of, i. 82.

Agassiz, his recognition of the scales of fishes in coprolites, i. 191 ; on causes of the death of fishes, i. 122 ; on origin of cololites, i. 200 ; on Glaris turtle, i. 257 ; his classification of fishes, i. 268 ; documents consigned to him by Cuvier, i. 267 ; his new orders of fishes, i. 269, 270 ; geological results established by, i. 272, 273 ; his new arrangement of Monte Bolca fishes, i. 285 ; his discovery of belemnites with ink bags, i. 374 ; on the bilateral structure of radiated animals, i. 415.

Agnostus, a genus of trilobites, i. 391.

Aichstadt, pterodactyles found at, i. 221.

Aix, fossil fishes of, i. 285.

Allan, Mr., his paper on Antrim belemnites, i. 377.

Amber, fossil resin from lignite, i. 520.

Amblypterus, fossil genus of fishes, i. 278.

Ammonites, formed by cephalopodous mollusks, i. 333 ; characteristic of different formations, i. 333 ; geological distribution of, i. 334 ; geographical distribution of, i. 335 ; extent and number of species, i. 334 ; size of, i. 334 ; sub-genera of, i. 334 ; shell composed of three parts, i. 336 ; external shells, i. 337, 338 ; outer chamber contained the animal, i. 337—339 ; double functions of shell, i. 338 ; contrivances to strengthen shells, i. 339—344 ; ribs, architectural disposition of, i. 341 ; transverse plates, use of their foliated edges, i. 344—349 ; compound internal arches, i. 349 ; siphuncle, organ of hydraulic adjustment, i. 350, 351 ; siphuncle, occasional state of preservation, i. 351, 352 ; siphuncle, placed differently from that of nautilus, i. 353 ; siphuncle, Dr. Prout's analysis of, i. 352 ; air chambers, more complex in ammonites than in nautili, i. 350 ;

ammonites, how different from nautili, 353 ; Von Buch's theory of, i. 352 ; uses of lobes and saddles in, i. 354 ; concluding observations upon, i. 355—357 ; probable place of heart in, ii. 58.

Anarrhicas, palatal teeth of, i. 281.

Animals, final cause of their creation, i. 101 ; lower classes of, predominate in earlier strata, i. 115 ; extinct races, how connected with existing species, i. 581 ; causes of their sudden destruction, i. 122 ; small number adapted for domestication, i. 100 ; terrestrial, how buried in strata of fresh water and marine formation, i. 128.

Animal enjoyment, one great object of creation, i. 293, 301.

Animal kingdom, four great divisions of, coeval, i. 61 ; early relations of, i. 87.

Animal life, extent of upon our globe, i. 102 ; progressive stages of, i. 115 ; remains of in secondary strata, i. 72.

Animal remains, most instructive evidences in geology, i. 128 ; preserved chiefly by agency of water, i. 126.

Annelidans, fossil remains of, i. 387.

Anning, Miss, her discovery of ink bag within horny sheath of belemnite, i. 374 ; her discoveries at Lyme Regis, passim ; her observations on connection of lignite with pentacrinites near Lyme, i. 437 ; her discovery of fossil pens and ink bags of Loligo, i. 304.

Anoplotherium, character and place of, i. 81.

Anstice, Mr. W., his discovery of insects in coal formation, i. 405 ; megalichthys, &c. found in Coalbrook Dale, by, ii. 43.

Ant eater, humerus like that of megatherium, i. 154.

Anthracotherium, character and place of, i. 82.

Apiocrinites, or pear encrinite, Miller's restoration of, i. 428.

Arachnidans, two great families of, found fossil, i. 405.

Arago, M., on expenditure of rain water, i. 557 ; on Artesian wells in France, i. 564.

indicate the food of ichthyosauri, and character of their intestinal canal, i. 197; derived from fishes in various formations, i. 198; polished for ornamental purposes, i. 199; conclusions from discovery of, i. 202; in coal formation near Edinburgh, i. 275; preserved in body of macropoma, i. 284.

Coral, secreted by polypes, i. 442; reefs, i. 444; their influence in the formation of strata, i. 445; fossil, inference from their state, i. 116; rag, extent of, in counties of Oxon, Bucks, Wilts, and Yorkshire, i. 445.

Corn-cockle muir, tracks of tortoises at, i. 259.

Cornwall, amount of steam power employed in, i. 534; invasions of by drifted sand, i. 127; disposition of metallic veins in, i. 550.

Corydalis, wing of, found in iron stone, of the coal formation, i. 410; ii. 77.

Cosmogony, Mosaic, the author's interpretation of, i. 20.

Cotta on fossil arborescent ferns, i. 465.

Crag, in Norfolk, geological place of, i. 179.

Craters, various phenomena of, ii. 8.

Creation, Mosaic account of, accords with natural phenomena, i. 13; origin of material elements by, i. 35.

Creator, necessity of, shewn by geology, i. 59.

Crinoideans, geological importance of, i. 416, 430; nature and character of, i. 417; most remarkable genera of, i. 417; living species rare, i. 418; abundance and importance of fossil species, i. 419, 430; anatomical structure of, i. 420; reproductive powers of, i. 421; early extinction of many species and genera, i. 430.

Crocodileans, fossil forms of, i. 249; slender character of snout, i. 250; habit probably piscivorous, i. 250.

Crocodiles, modern, habits of, i. 250; gavial, gangetic, piscivorous, i. 250; functions of fossil species, i. 251; Cuvier's observations on, i. 252; number of living and fossil species of, i. 252; dentition, provisions in mode of, i. 254; fossil forms of, at variance with all theories of gradual transmutation or developement, i. 254.

Crustaceans, extent of fossil remains, of, i. 387.

Crystalline rocks, influenced by chemical and electro-magnetic forces, i. 36; eight distinct varieties of, i. 37; their position beneath stratified rocks, i. 42; probable igneous origin of, i. 39; gradations in character of, i. 41; proofs of intention in phenomena of, i. 45; proofs of design afforded by, i. 574.

Crystals, definite forms and composition of, i. 575—577; component molecules of, i. 574, 577.

Ctenoidean order of fishes, i. 270.

Curculionidæ in iron stone of Coalbrook Dale, i. 409.

Cuttle fish, structure and habits of, i. 303; internal ink bag of, i. 303.

Cuvier, his conclusion that organic life has not existed from eternity, i. 59; his account of the basin of Paris, i. 76; his account of discoveries at Mont Martre, i. 83; consigns his materials for a work on fossil fishes to M. Agassiz, i. 200, 267; his conjecture concerning plesiosaurus, i. 208; had observed nearly 8,000 species of living fishes, i. 265; perfection of his reasoning on contrivances and compensations in the structure of animals, i. 140.

Cycadeæ, abundant in strata of the secondary series, i. 490, 491; number and extent of recent and fossil species, i. 491; leaves fossil in oolite of Yorkshire and at Stonesfield, i. 492; in coal formation of Bohemia, i. 492; habit and structure of, i. 492; intermediate character of, i. 493; fossil on the coast of Dorset, i. 494; peculiarities in structure of trunk of, i. 494, 496; mode of increase by buds, i. 499; link supplied by the discovery of, i. 502.

Cycadites, once natives of England, i. 495; microphyllus, microscopic structure of, 497—501; megalophyllus, buds in axillæ of scales, i. 500; resemblance of fossil and living species, i. 501.

Cycas revoluta, buds on trunk of, i. 499; circinalis, height of, i. 494.

Cycloidean order of fishes, i. 270.

Cypris, microscopic shells of, in Wealden formation, i. 118; in coal formation near Edinburgh, i. 275.

DAPEDIUM, scales of, i. 282.

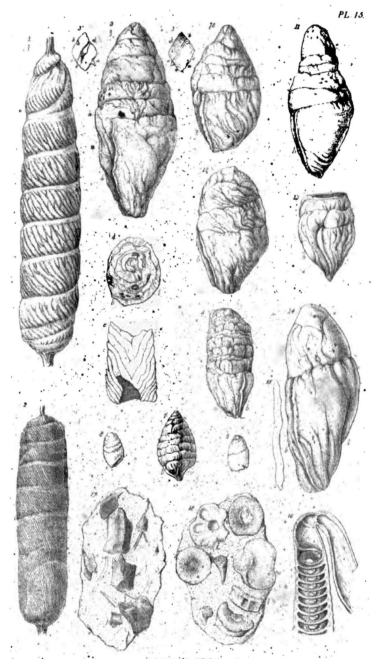

COPROLITES.

Chiefly from the Lias at Lyme Regis.

COLOLITE,
in the Lithographic Slate of Solenhofen
i Humanidil Lithography

Pl. 16.

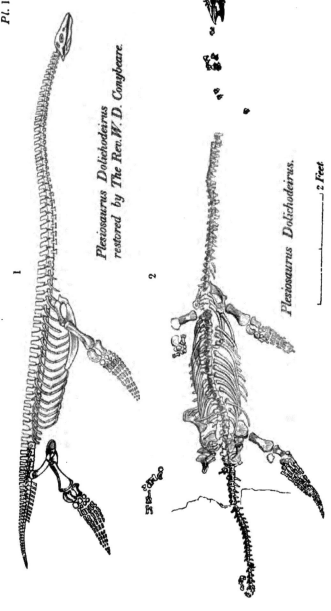

1

Plesiosaurus Dolichodeirus restored by The Rev. W. D. Conybeare.

2

Plesiosaurus Dolichodeirus.

————— 2 *Feet*

Pl. 17.

Plesiosaurus Dolichodeirus.

1 Foot

Pl. 18.

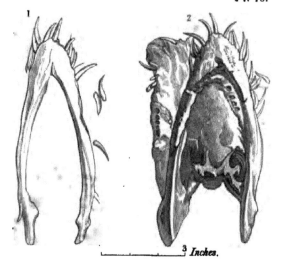

Head and Jaws of Plesiosaurus Dolichodeirus.

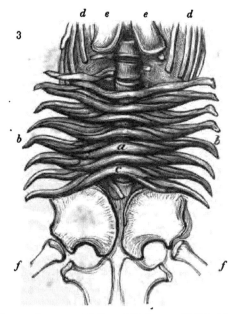

Compound sterno costal Arcs of Plesiosaurus Dolichodeirus.

Pl. 19.

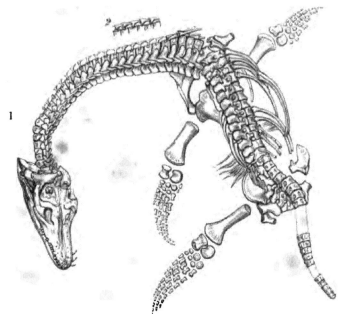

Plesiosaurus Macrocephalus. |_____| 6 Inches.

3

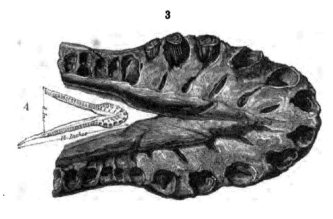

Pl. 20.

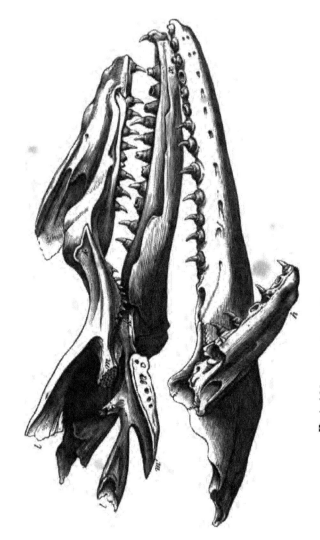

Head of Mosasaurus from the Upper Cretaceous formation at Maestricht.

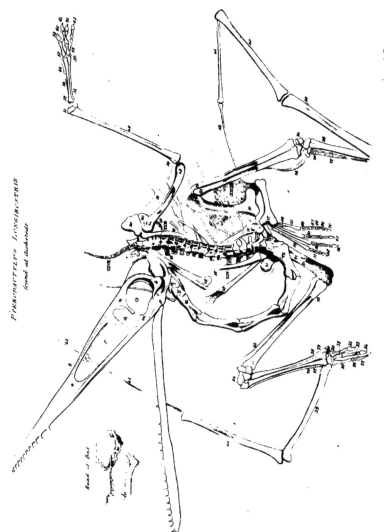

PTERODACTYLUS LONGIROSTRIS
found at Aichstedt

Head of Bat

J. Fisher sc.

Pl. 23.

Right Lower Jaw of Megalosaurus from Stonesfield Oxon.

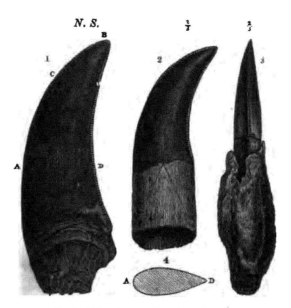

Teeth of Megalosaurus from Stonesfield, Oxon.

Pl. 24.

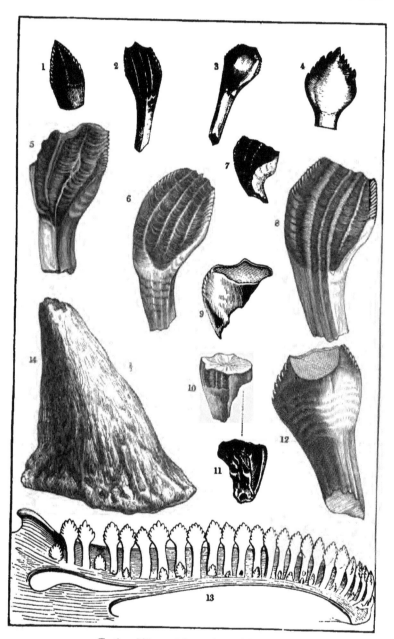

Teeth and Bones of Iguanodon and Iguana.

Pl. 25.

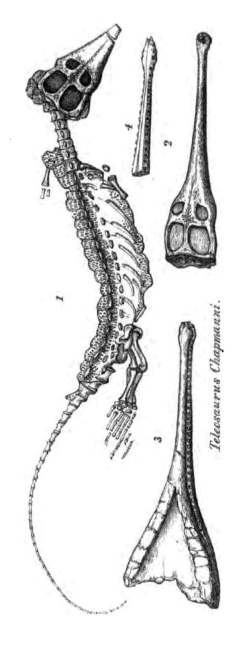

Teleosaurus Chapmanni.

Crocodilian remains from the Lias of Whitby and Oolite of Oxfordshire.

1

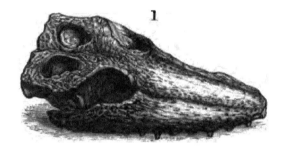

Crocodilus Spenceri from I. of Sheppey.

2   3

2. *Teleosaurus.*   3. *Steneosaurus.*

4

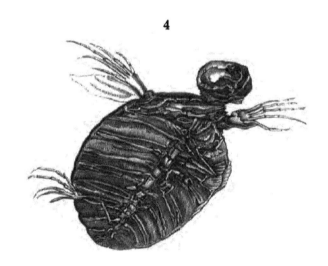

Impressions of Footsteps on red Sandstone near Dumfries.

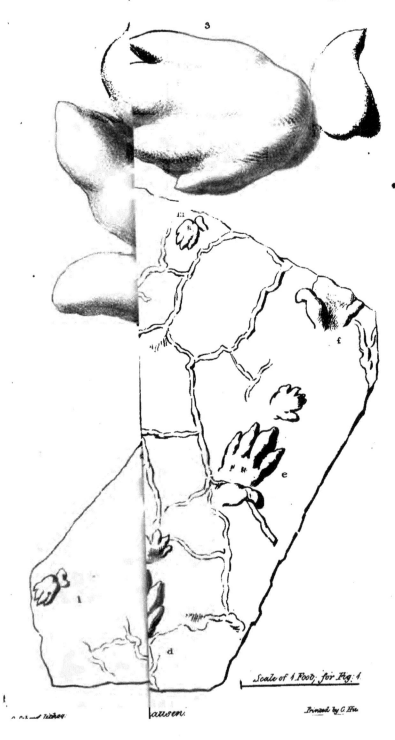

Pl 26

Scale of 1 Foot; for Fig. 1.

Nat. Size

8 Inches space between

Nat. Size.

Footsteps of some unknown Reptile
on a slab of New-red sandstone.
from Hessberg near Hildburghausen.

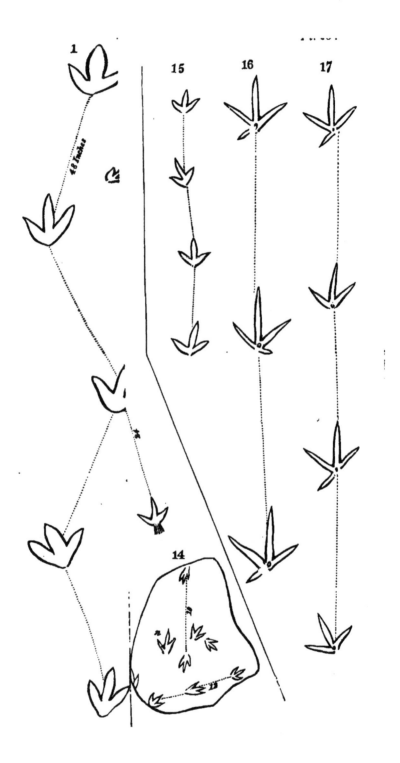

Pl. 27.

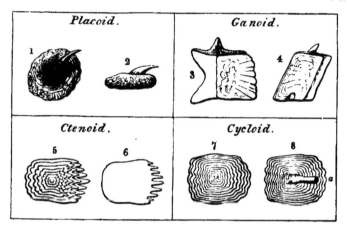

Characteristic scales of the four orders of Fishes.—Agassiz.

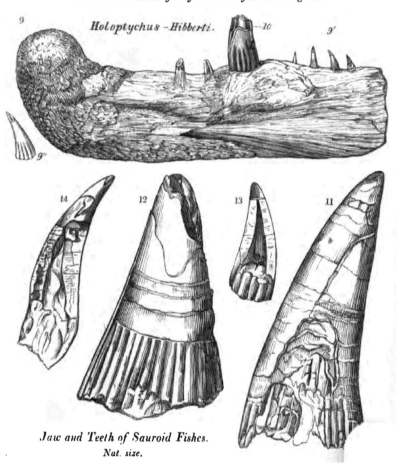

Jaw and Teeth of Sauroid Fishes.
Nat. size.

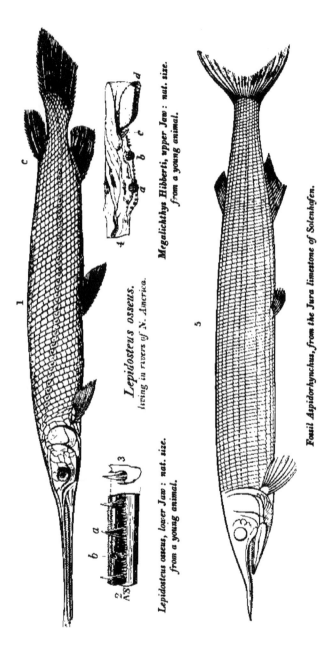

Lepidosteus osseus.
living in rivers of N. America.

Lepidosteus osseus, lower Jaw: nat. size.
from a young animal.

Megalichthys Hibberti, upper Jaw: nat. size.
from a young animal.

Fossil Aspidorhynchus, from the Jura limestone of Solenhofen.

Examples of recent and fossil Sauroid Fishes.

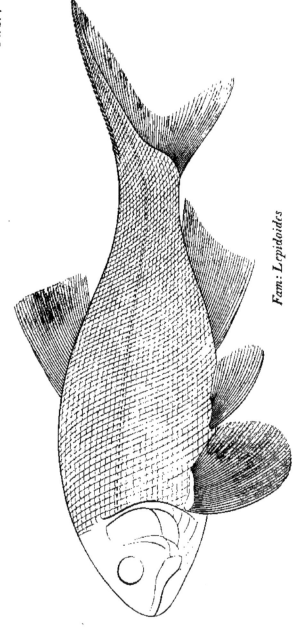

Fam: *Lepidoides*

Amblypterus from the Coal formation at Saarbrück. Half nat. size.

Pl 37.

1. *Microdon.*
2. *Gyrodus palatal teeth.*
3. *Pycnodus palatal teeth.*

Fam: Pycnodontes.

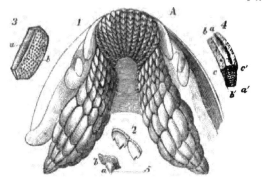

Teeth of Port Jackson Shark. *Cestracion Phillippi.*

B

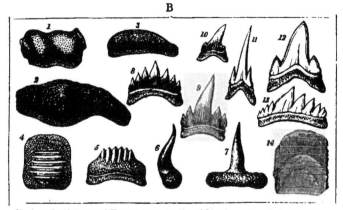

Various fossil forms of Teeth in the family of Sharks. 14. *Palata of Ray.*

C

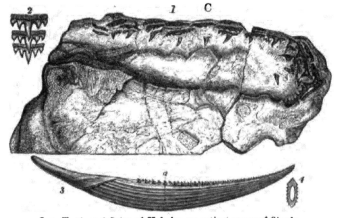

Jaw, Teeth, and Spine of Hybodus, an extinct genus of Sharks.

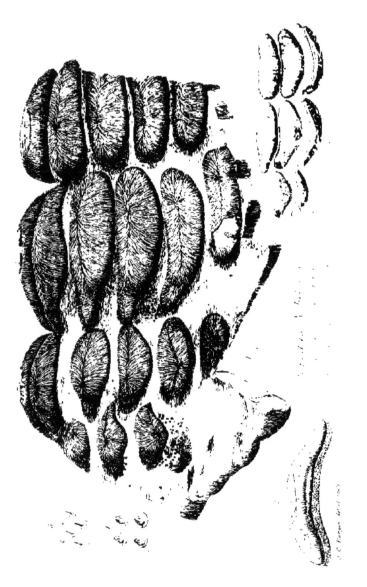

Tab. 27 f

S.C.Burgon del.

Ptychodus Polygyrus. Agass.

In the Cabinet of J.J.Burgon Esq.

J.Erxleben sculp.

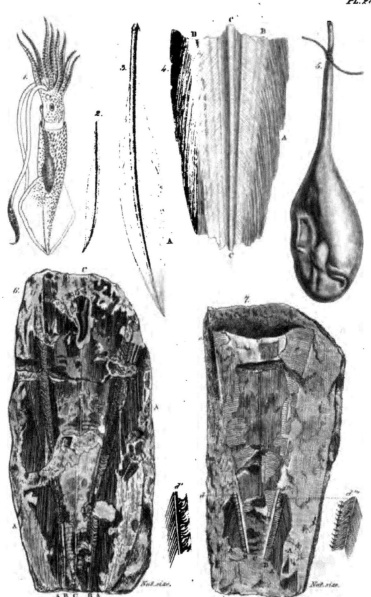

PENS OF RECENT LOLIGO, INK BAG OF SEPIA OFFICINALIS
AND FOSSIL PENS FROM THE LIAS AT LYME REGIS.

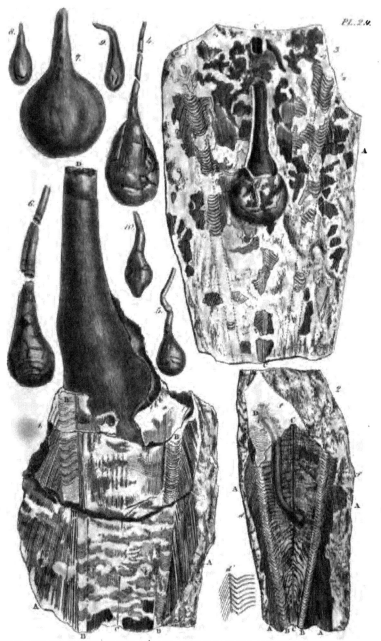

FOSSIL PENS AND INK BAGS OF LOLIGO FROM LIAS &

Natural size.

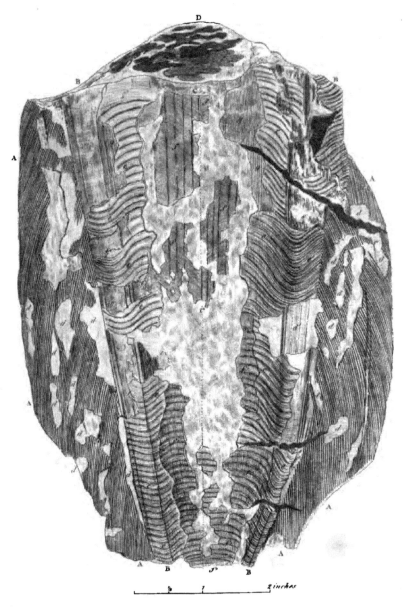

CONVEX SURFACE OF A FOSSIL PEN OF LOLIGO FROM LIAS AT LY

Shewing the Structure of the component Laminæ.

Nautilus pompilius

RHYNCHOLITES.

Pl. 32.

1 & 2. NAUTILUS HEXAGONUS

NAUTILUS STRIATUS,

in Lias at Whitby.

Zeitter. sc.

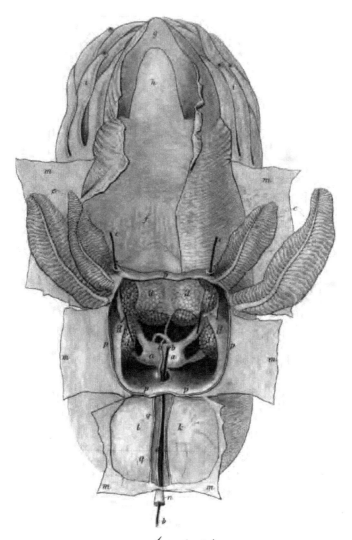

Animal of the
NAUTILUS POMPILIUS.

n. del.

Zeitter, s

AMMONITES OBTUSUS.

From Lias, at Lyme Regis.

r. del.

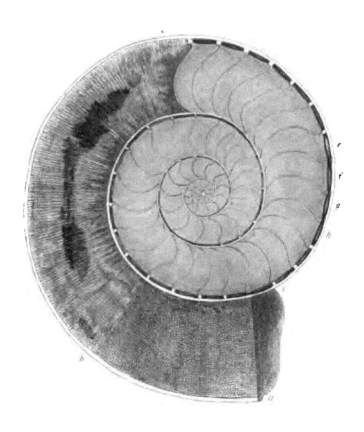

Longitudinal Section of
AMMONITES OBTUSUS.

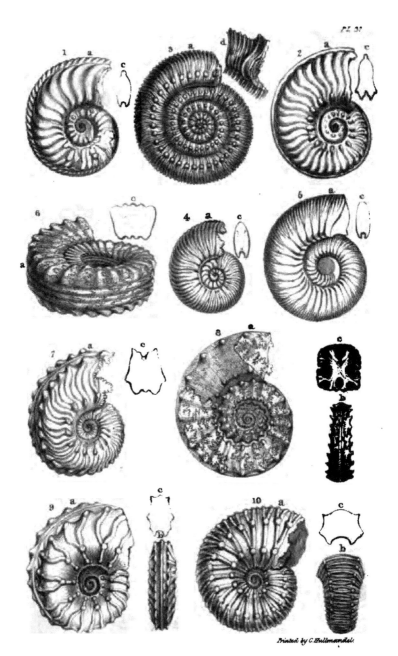

Various Forms of Ammonites.

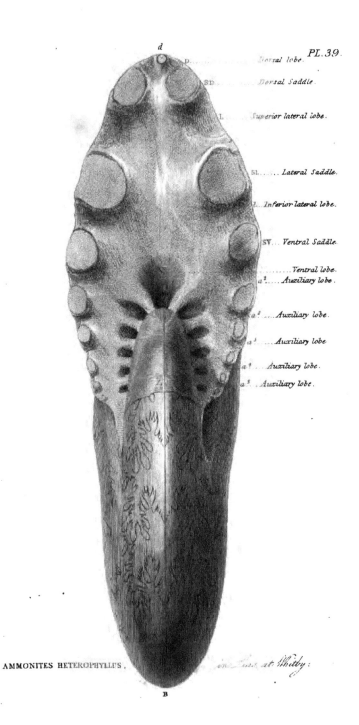

d　PL. 39.

Dorsal lobe.

p..................

SD.........Dorsal Saddle.

L.........Superior lateral lobe.

SL.......Lateral Saddle.

l....Inferior lateral lobe.

SV....Ventral Saddle.

.............Ventral lobe.
a²......Auxiliary lobe.

a³.....Auxiliary lobe.

a⁴......Auxiliary lobe

a⁵....Auxiliary lobe.

a⁶....Auxiliary lobe.

AMMONITES HETEROPHYLLUS.　on Lias at Whitby:

B

Fisher del.

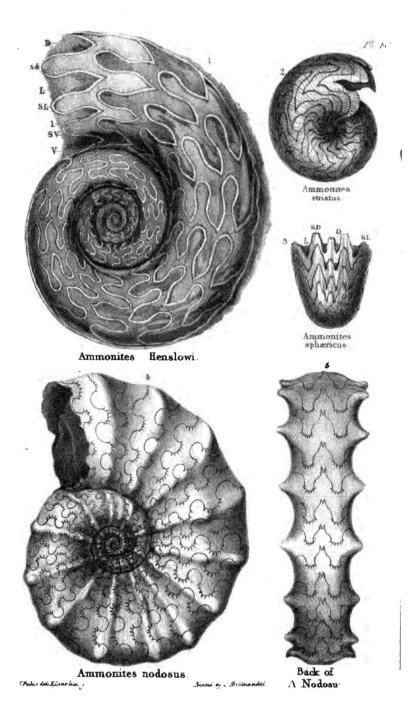

Pl. 1.

D
s.d
L
S.L.
I
S.V
V

1

2

Ammonites
striatus.

s.D   D
3   L   S.L.

Ammonites
sphæricus

Ammonites   Henslowi.

4

6

Ammonites nodosus.

Back of
A Nodosu.

T.Pake del.Kilear lith.   Printed by C. Hullmandel

Pl.41.

Winding Partitions between the Air Chambers of
AMMONITES GIGANTEUS.

C. Hullmandel's Lithography

Pl. 42.

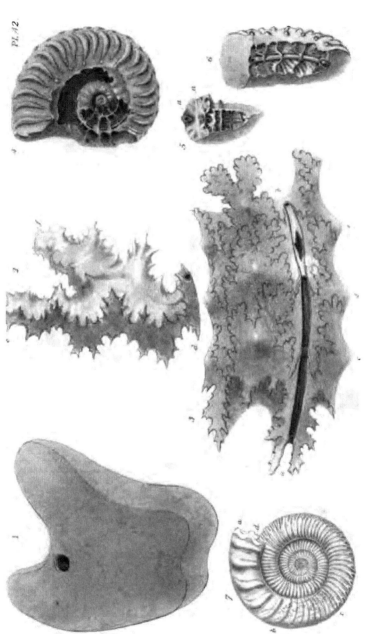

1. NAUTILUS HEXAGONUS. 2. AMMONITES EXCAVATUS. 3. AMMONITES CATENA. 4.5.6. AMMONITES VARICOSUS. 5.6. Fragments of fig. 4.
7. AMMONITES VARIOCOSTATUS.

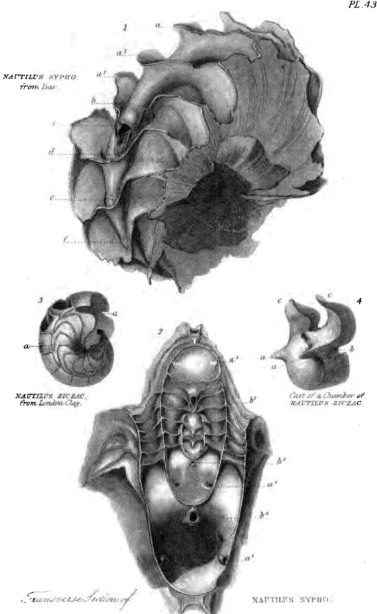

NAUTILUS SYPHO.
from Das.

NAUTILUS ZIC ZAC,
from London Clay.

Cast of a Chamber of
NAUTILUS ZICZAC.

Transverse Section of   NAUTILUS SYPHO.

Fisher, del.

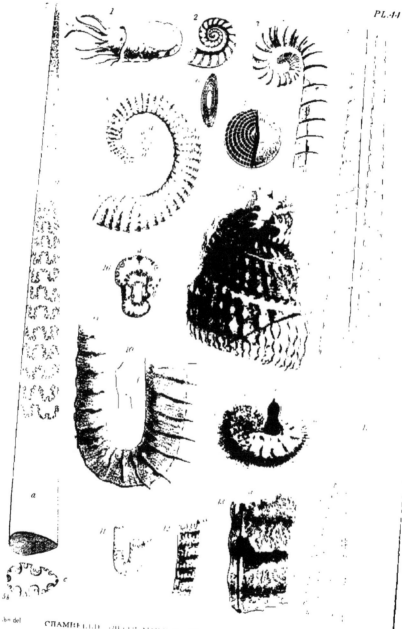

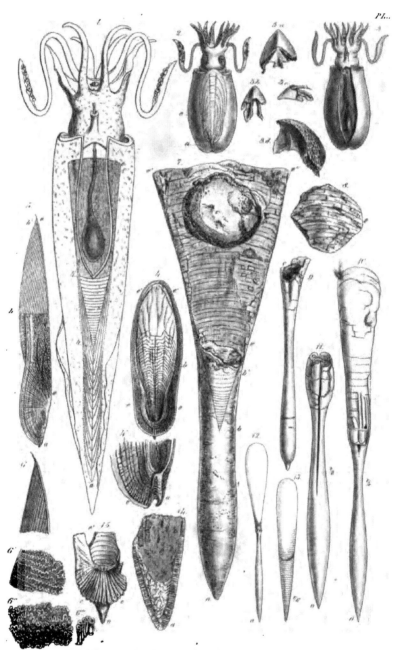

ILLUSTRATIONS of the GENUS BELEMNOSEPIA.

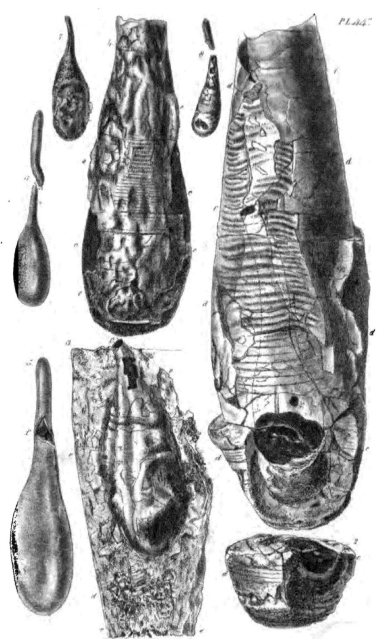

INK BAGS OF BELEMNOSEPIA
in their NACREOUS SHEATHS, from the LIAS of LYME R.

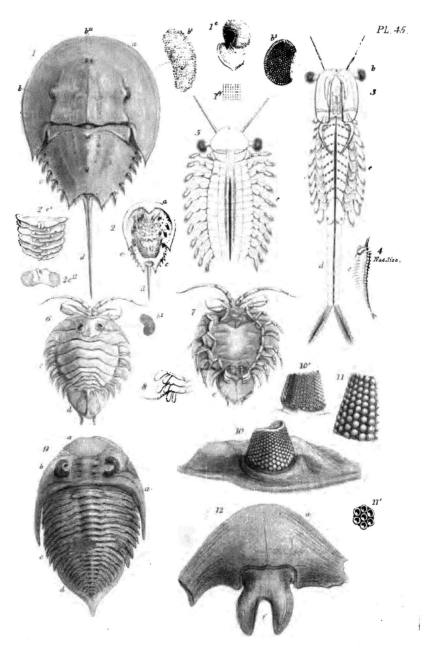

PL. 45.

L.Fisher del.   TRILOBITES & recent Animals allied to them.

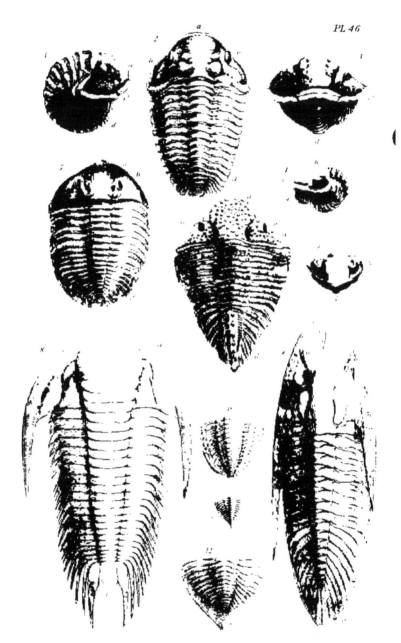

sher del.

TRILOBITES.

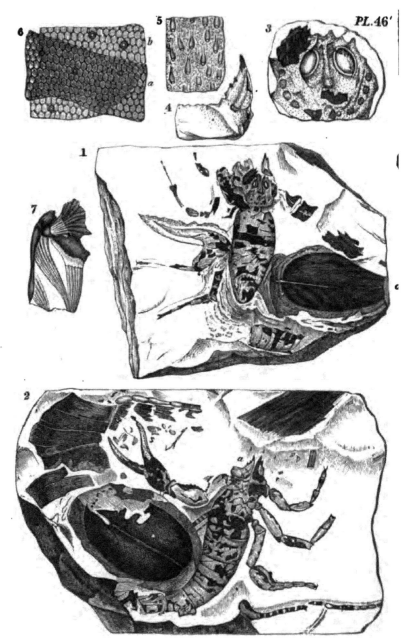

Fossil Scorpion from the Coal Formation at Chomle in Bohemia.

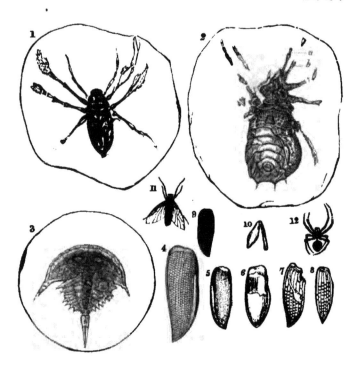

Fossil Insects, Spider, and Limulus.   Nat. size.

Tail and Intestine of Scorpion from Chomle in Bohemia.

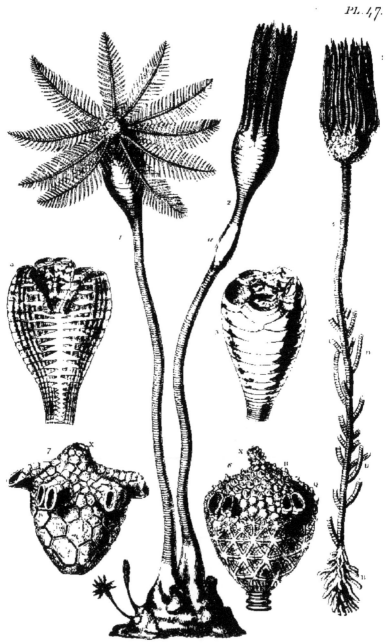

APIOCRINITES & ACTINOCRINITES.

Pl. 48.

Fragment of a Lily Encrinite.   Encrinites Moniliformis.

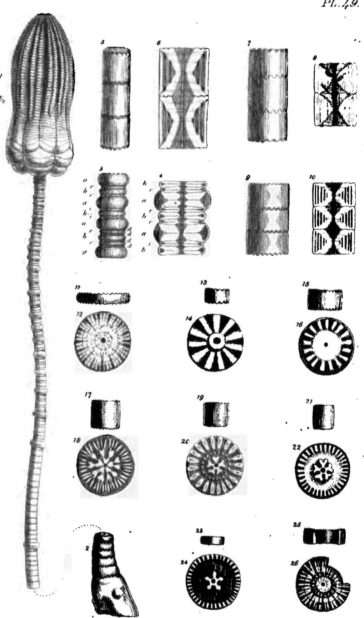

ENCRINITES MONILIFORMIS.

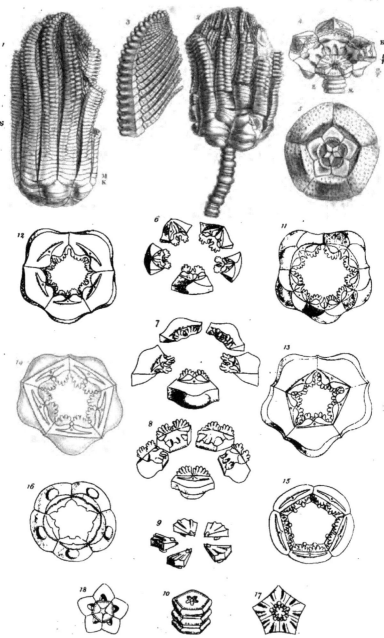

Pl. 50

ENCRINITES MONILIFORMIS.

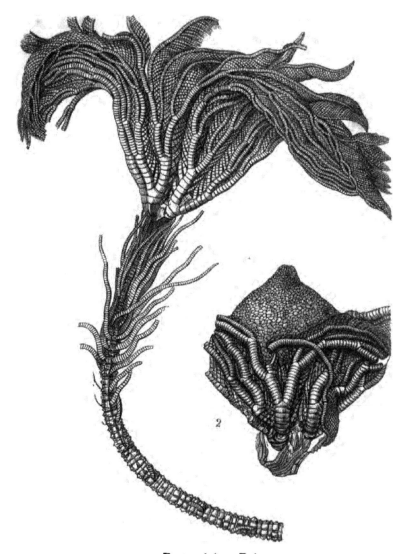

Pentacrinites Briareus.

From the Lias at Lyme Regis.   1. *Nat. size.*—2. $\frac{2}{1}$

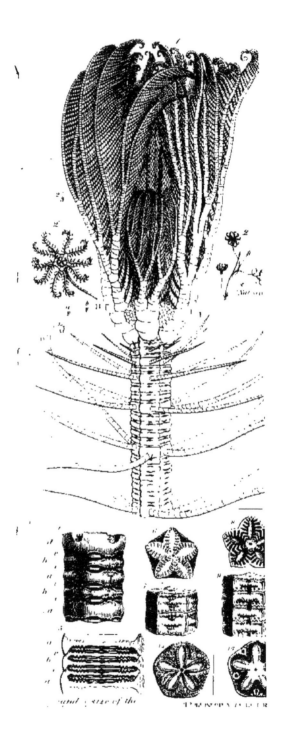

Pl. 58

PENTACRINITES BRIAREUS, FROM THE LIAS AT LYME REGIS.DORSET, IN THE COLLECTION OF PROFESSOR SEDGWICK

Pl. 54.

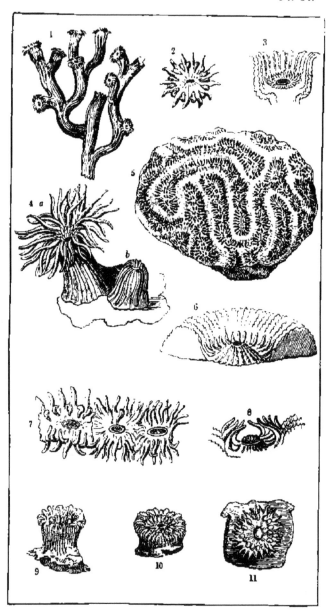

Recent Corals with their Polypes.

Pl. 55.

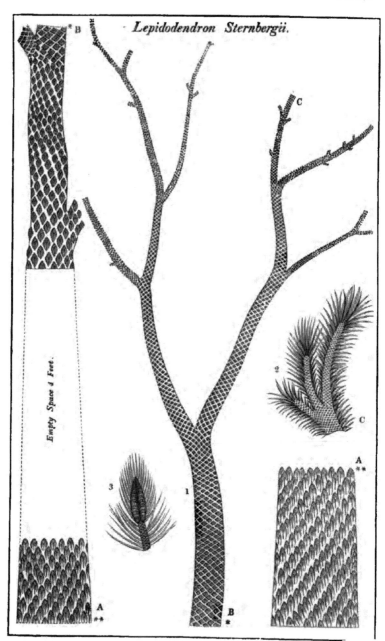

Lepidodendron Sternbergii.

Fossil Tree found prostrate in a Coal Mine at Swina in Bohemia.

Pl. 56.

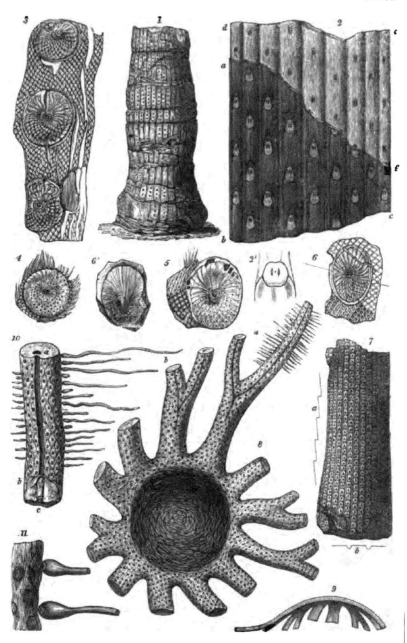

Remains of Plants, of extinct Families, from the Coal Formation.

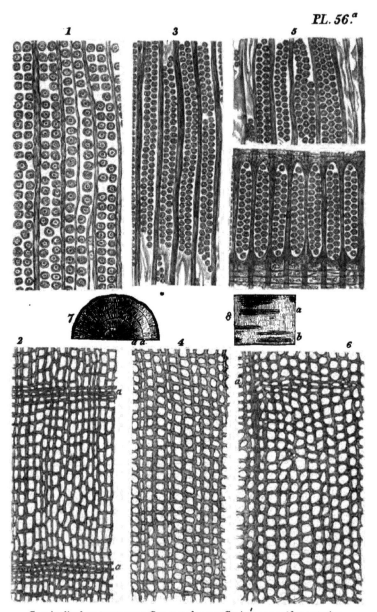

Longitudinal and transverse Sections of recent Coniferæ magnified 400 times.

Pl. 57.

Fig. 1.

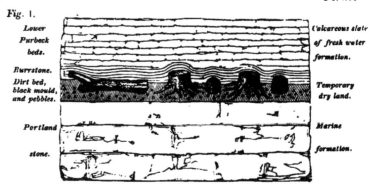

Lower
Purbeck
beds.

Burrstone.
Dirt bed,
black mould,
and pebbles.

Portland

stone.

Calcareous slate
of fresh water

formation.

Temporary
dry land.

Marine

formation.

Section of the Dirt-bed in the Isle of Portland, shewing the subterranean remains
of an ancient Forest.   De la Beche.

Fig. 2.

Circular ridges and depressions on the Burrstone, in the I. of Portland, round an erect
stump of a tree, four feet high.   Sketched by Professor Henslow, 1832.

Fig. 3.

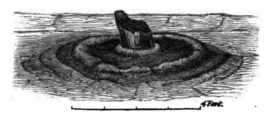

Lower Purbeck beds
composed of calcare-
ous slate of freshwater
formation.

Soft burrstone.
Ancient forest in the
dirt bed.

Portland stone of
marine formation.

Section of the Cliff east of Lulworth Cove, Dorset.   Bucklund.

Pl. 58.

Cycas Revoluta with buds proceeding from the Axillæ of the bases of fallen leaves. Scale ¼.

Pl. 59.

Zamia Pungens with its fruit.

Transverse section of Zamia
Horrida.   Scale ⅐.

Transverse section of a young
Trunk of Cycas Revoluta. ½.

Pl. 60.

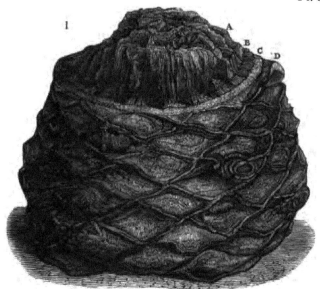

Trunk of Cycadites Megalophyllus, from I. of Portland. Scale ½.

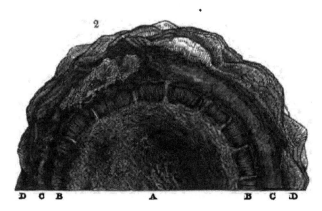

Transverse section of the Trunk of Cycadites Megalophyllus from I. Portland. Scale ½.

Pl. 61.

Fig. 1.

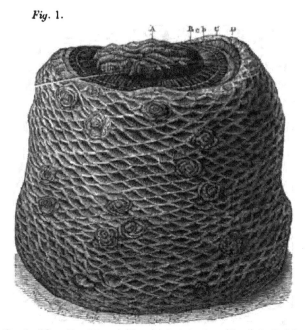

Trunk of Cycadites Microphyllus with buds in the axillæ of the Petioles
from I. Portland.   Scale ½.

Fig. 2.   *Fig.* 3.

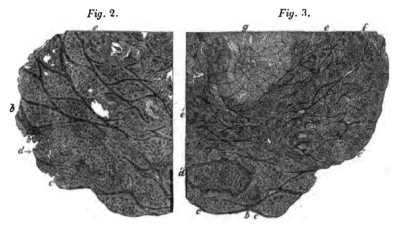

Sections of Buds and Petioles of Cycadites Microphyllus from I. Portland.   Nat. size.

Pl. 62.

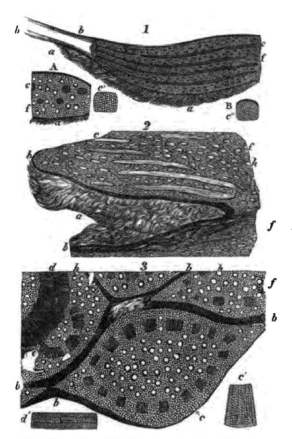

Magnified sections of Petioles of recent and fossil Cycadeæ.

Pl. 64.

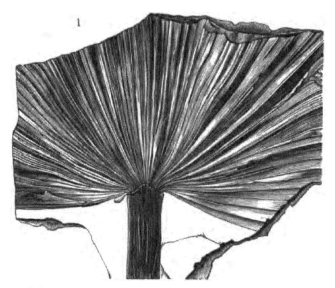

1

Palmacites Lamanonis from the Tertiary Gypsum at Aix.   Scale $\frac{1}{4}$.

2

Endogenites Echinatus from the Calcaire grossier near Soissons.   Scale $\frac{1}{6}$.

Pl. 65.

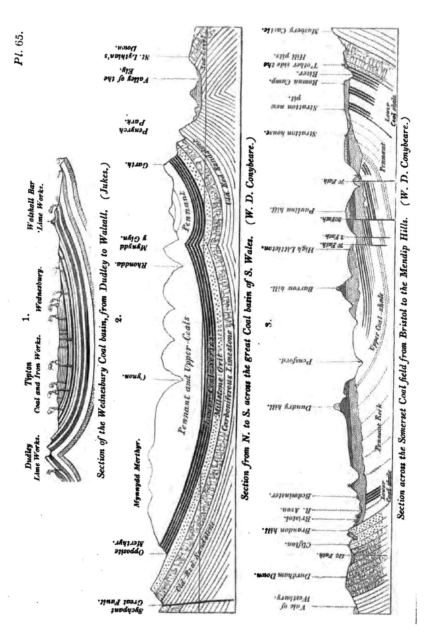

Dudley
Lime Works.

Tipton
Coal and Iron Works.

1.

Wednesbury.

Walsall Bar
Lime Works.

Section of the Wednesbury Coal basin, from Dudley to Walsall. (Jukes.)

2.

Myrnydd Merthyr.

Opposite
Merthyr.

Sychpant
Great Fault.

(Cynon.)

Rhondda.

Mynydd
y Glyn.

Garth.

Penyrch
Park.

Valley of the
Ely.

St. Lythian's
Down.

Section from N. to S. across the great Coal basin of S. Wales. (W. D. Conybeare.)

3.

Vale of
Westbury.

Durdham Down.

110 Fault.

Clifton.

Brandon hill.

Bristol.
R. Avon.

Bedminster.

Pensford.

Dundry hill.

Barrow hill.

High Littleton.

Pulton hill.

Stratton house.

Stratton new
pit.

Roman Camp.

'T'other side the
River.'

Hill pits.

Maesbery Castle.

Section across the Somerset Coal field from Bristol to the Mendip Hills. (W. D. Conybeare.)

Pl. 66.

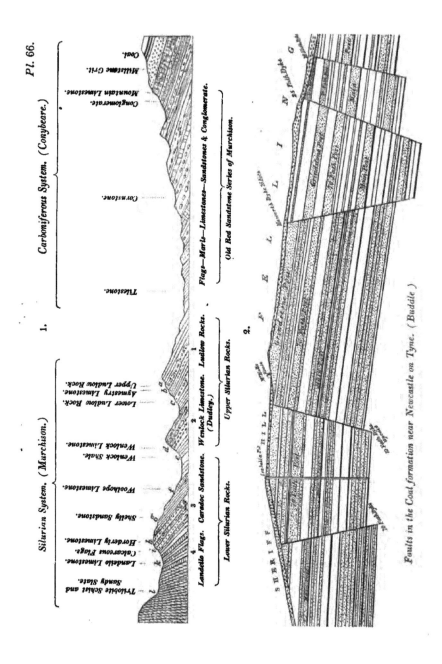

Silurian System. (Murchison.) Carboniferous System. (Conybeare.)

1.

Trilobite Schist and Sandy Slate.
Landeilo Limestone.
Calcareous Flags.
Horderly Limestone.
Shelly Sandstone.
Woolhope Limestone.
Wenlock Shale.
Wenlock Limestone.
Lower Ludlow Rock.
Aymestry Limestone.
Upper Ludlow Rock.

Tilestone.
Cornstone.
Conglomerate.
Mountain Limestone.
Millstone Grit.
Coal.

4 Landeilo Flags. Caradoc Sandstone. 3 Wenlock Limestone. Ludlow Rocks. 2 (Dudley.) 1

Lower Silurian Rocks. Upper Silurian Rocks. Old Red Sandstone Series of Murchison.

Flags—Marls—Limestones—Sandstone & Conglomerate.

2.

Faults in the Coal formation near Newcastle on Tyne. (Buddle.)

Pl. 67.

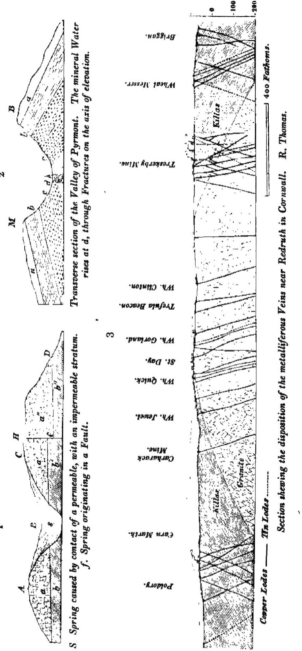

1

S Spring caused by contact of a permeable, with an impermeable stratum.
f. Spring originating in a Fault.

2

Transverse section of the Valley of Pyrmont. The mineral Water
rises at d, through Fractures on the axis of elevation.

3

Section shewing the disposition of the metalliferous Veins near Redruth in Cornwall. R. Thomas.

Copper Lodes ——— Tin Lodes ....

400 Fathoms.

Fathoms.
0
100
200

Pl. 68.

Section shewing the cause of the rise of water in Artesian Wells in the basin of London.

Leighton Buzzard. — Dunstable. — St. Albans. — Knockholt. — Sevenoaks. — Sydenham. — R. Thames. — London. — Hampstead.

London Clay

Plastic Clay

Chalk

Firestone

Gault Clay

Woburn Sand.